Display Creative Numeracy

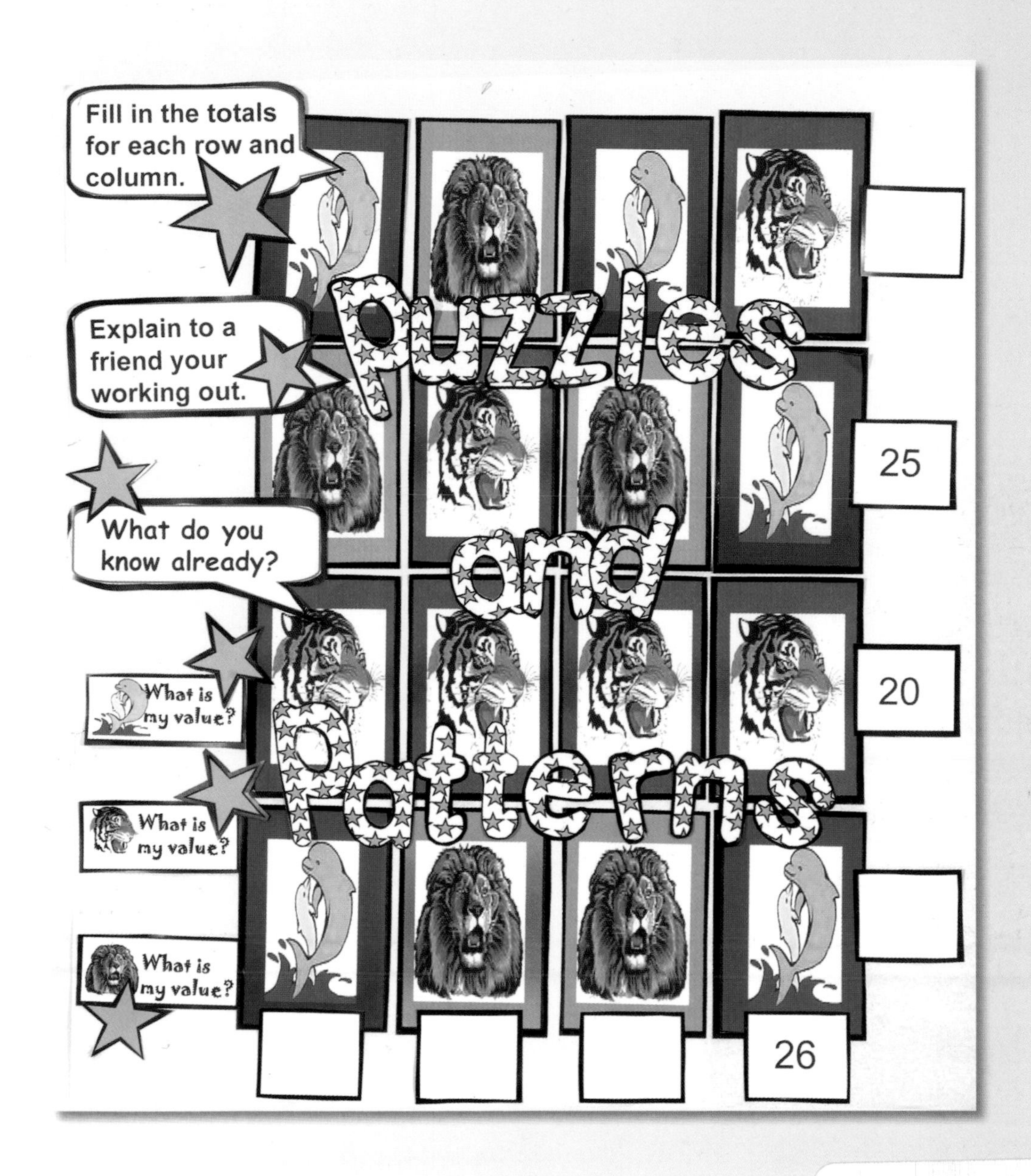

Margaret Share

Acknowledgements

The author would like to give a heartfelt thank you to everybody involved in the making of this book for their support, dedication and inspiration. Special thanks to Sheila Bean, head teacher at Tranmoor Primary School, Doncaster, and her staff and pupils; Olwen Hawkes, head teacher at Church Vale Primary School, Warsop, and her staff and pupils; Christine Cross, head teacher at Hallcroft Infants, Retford, and her staff and pupils; and to Zoe Nichols, Steve Forrest and Caroline Pook.

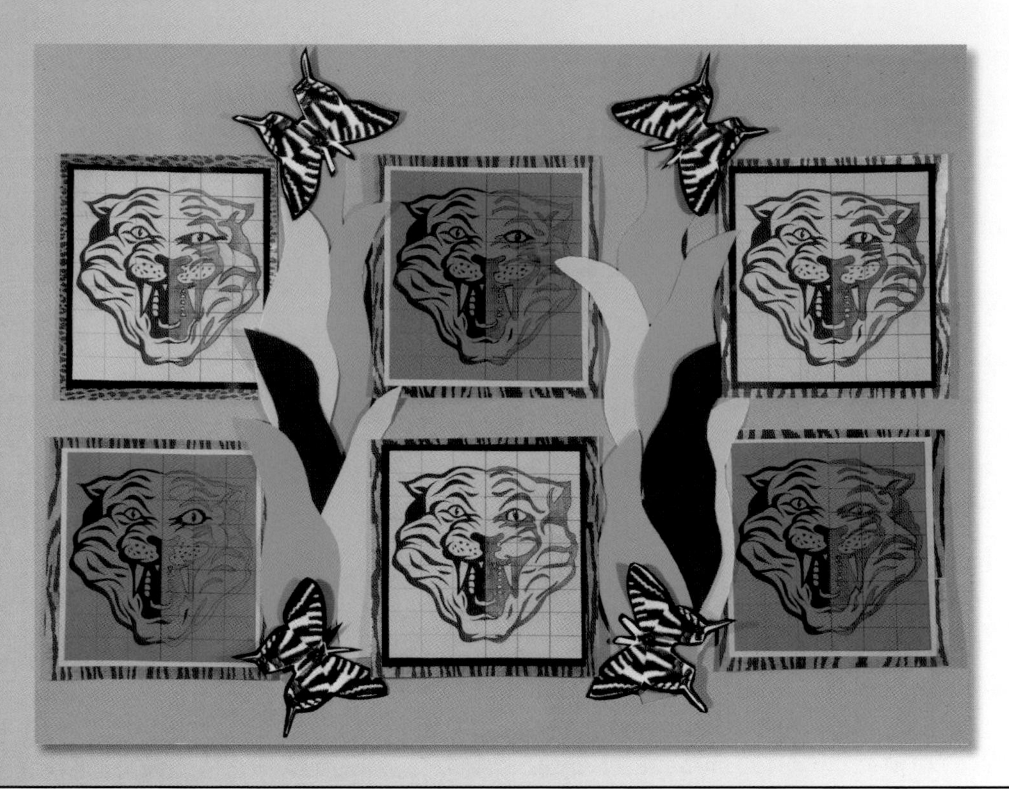

Belair Publications, Waterslade House, Thame Road, Haddenham, Buckinghamshire, HP17 8NT

Email: belair@belair-publications.co.uk

Commissioning Editor: Zoë Nichols
Page layout: Barbara Linton
Cover design: Steve West
Editor: Caroline Pook
Photography: RB Photography and Steve Forrest

First published in 2008 by Belair Publications.

British Library Cataloguing in Publication Data. A catalogue record for this publication is available from the British Library.

ISBN 978 1 84191 4602

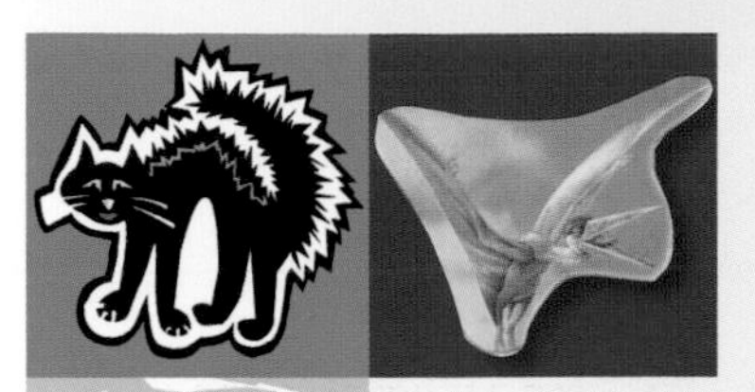

Contents

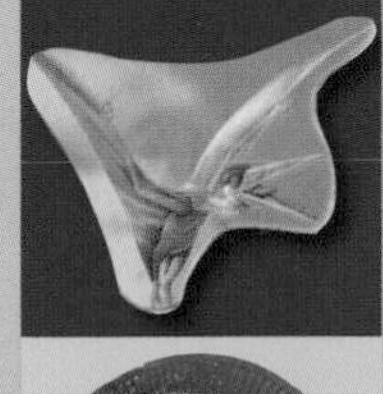
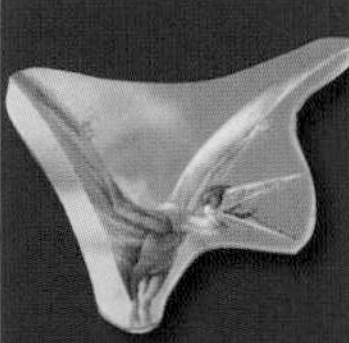

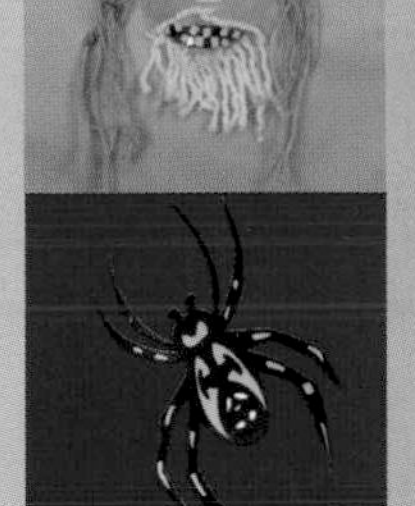

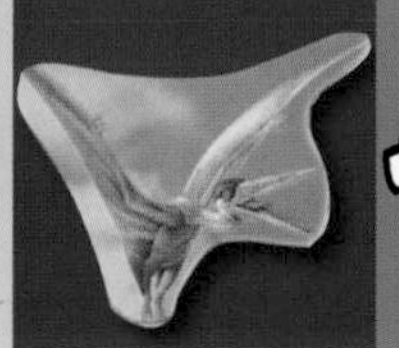

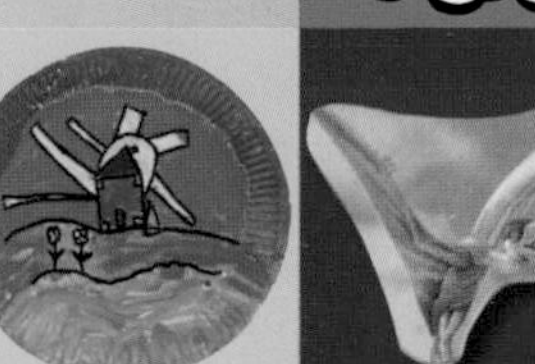

Introduction

Welcome to *Display Creative Numeracy*, which demonstrates how maths can be linked across the curriculum in practical and real-life situations, enhancing both the topic and the maths concept. The 33 themes of the book are divided into three main sections: Art and Display, Starting Points, and Working Walls.

Art and Display

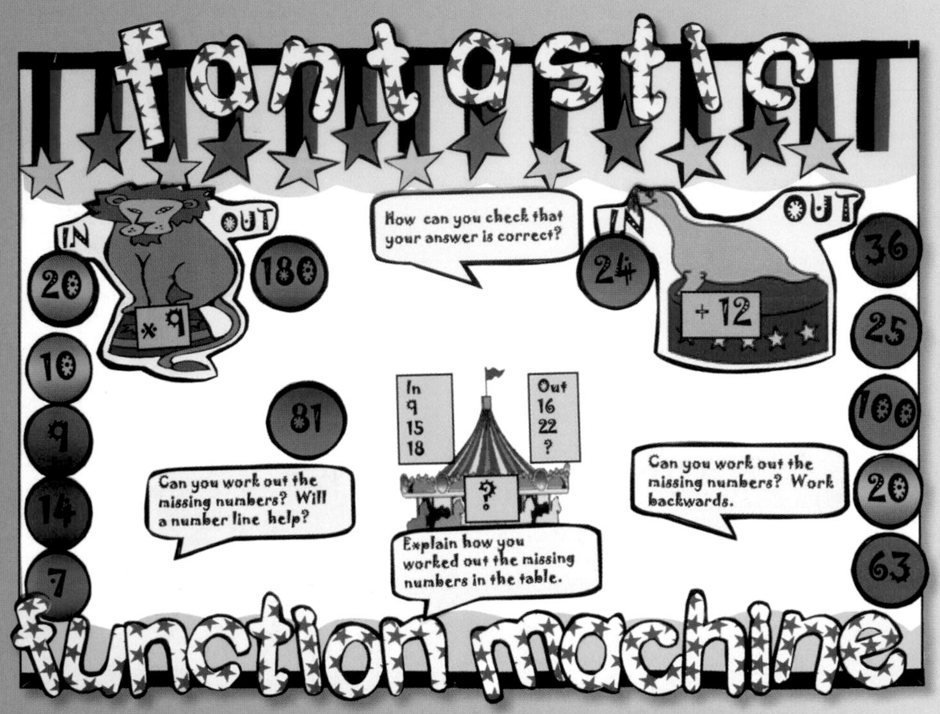

The Art and Display section describes how to make the displays and offers ideas for how to use them in an interactive way. They are quick to produce and support the learning and teaching of key objectives: therefore, they are not just attractive 'wallpaper' but an integral part of learning. The displays have been designed to encourage interactivity: some identify key points and suggestions for understanding challenging parts of a concept. The displays often include questions such as, 'How do you know?' and encourage children to use the word 'because' to support the structuring of their answers. Very clear models for calculations, including number lines, are often included to provide accurate prompts when calculating. Activities in each theme have been designed to support the key objectives for the display and the relevant vocabulary.

Starting Points

The Starting Points include many oral and mental activities that are designed to support kinaesthetic work. The use of whiteboard activities is an integral part of Starting Points: they are used to check children's understanding through 'Show me' activities. Activities are designed to encourage children to work in pairs and groups, and to develop their understanding through games and practical activities. Emphasis is placed on developing mathematical vocabulary within speaking and listening activities.

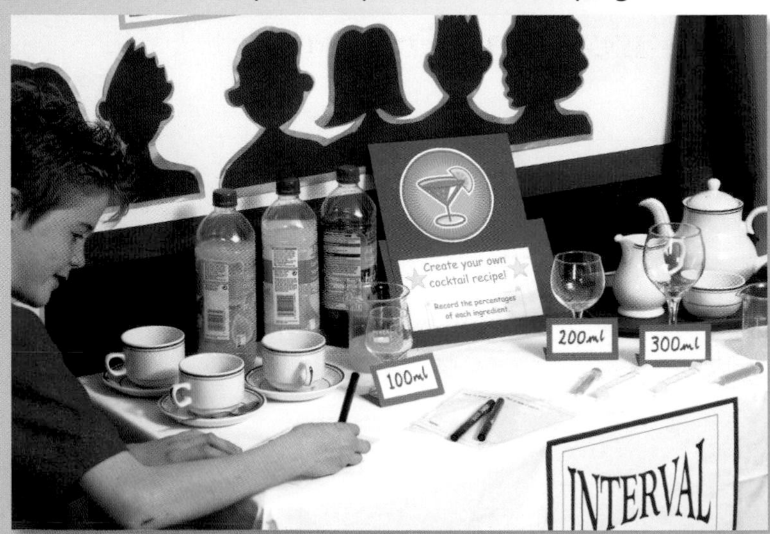

Further Activities

Further Activities provide children with the opportunity to extend the concepts introduced in Starting Points in a problem solving context. Table-top activities are integral to children's learning. The displays have been designed to support the learning of key objectives through games and practical activities, and children will also have the opportunity to use apparatus.

Role-play areas can be extremely effective and popular throughout the primary age range. Role play is often prioritised with younger children, but it is a powerful tool that older children can also use to engage with their own learning. It supports speaking and listening in maths and, in particular, encourages children to use the correct mathematical vocabulary and to use explanations in a practical situation. Ideas are given for role-play areas in some of the themes in this book: for example, in The Theatre on page 27.

Cross-curricular Links

Maths is meaningful when it is used and applied across the curriculum and used in real-life situations. For example, probability becomes more interesting when considered in relation to which team is likely to win a football match. Suggested cross-curricular links have therefore been included at the end of some of the themes.

Working Walls

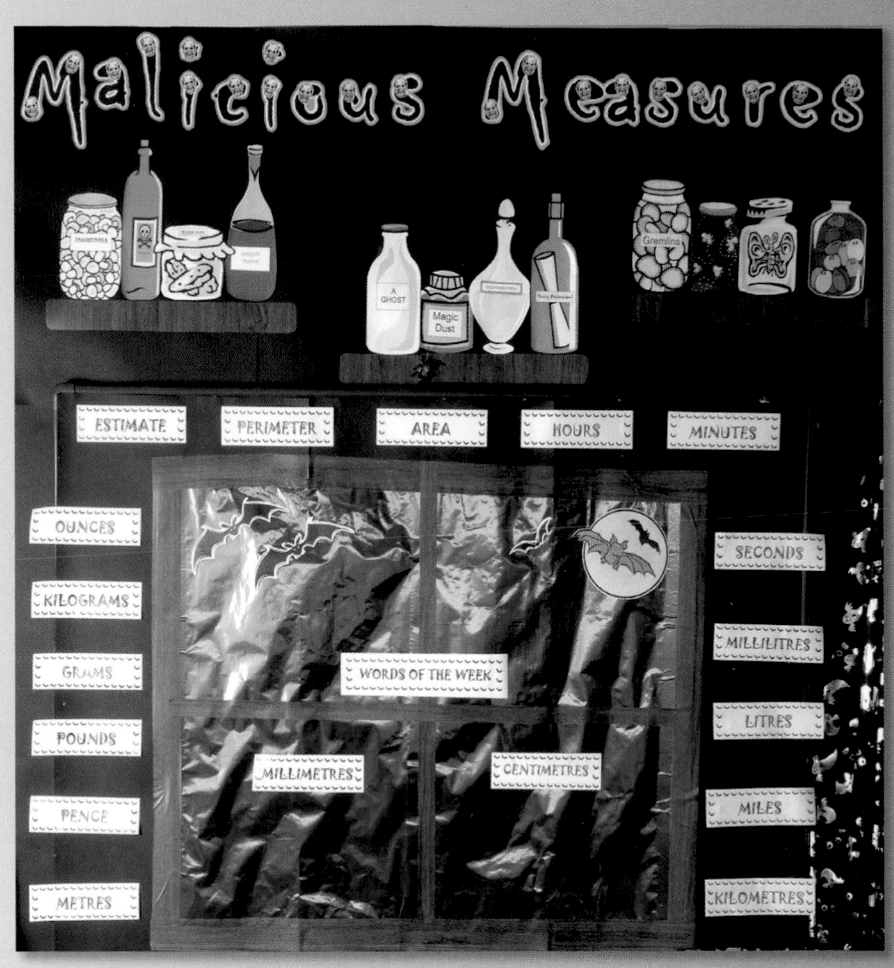

It is important that the classroom environment supports both the learning and teaching of maths. Displays have different purposes: those that are cross curricular will be exciting, colourful and interactive, and can be designed to remain for the duration of the topic, which could be several weeks. Use Blu Tack rather than staples, so that examples of charts, diagrams and written calculations can easily be changed as the topic progresses. These displays lift the status of maths in the classroom and encourage children to see that maths is real and relevant to them in their world.

Working walls are a different type of maths display. They are designed to be quick to create and are displayed for a single lesson or series of lessons. They support children with the current unit of work, focusing on the objective by displaying helpful diagrams, tables, written calculation models, vocabulary and models, using familiar apparatus. They encourage independence, in that children can use them as a 'prompt' during the lesson. You can refer to them as a demonstration model throughout the lesson and to recap on previous work. Working walls are explained in more detail on page 72.

Resources and Equipment

The use of ICT has transformed the way we teach maths; however, it is essential that children have the opportunity to use practical resources that support the formation of concepts. Wherever possible, ideas have been included that use quick and easy apparatus to support concept formation.

I hope this book will provide you with inspiration to look for connecting links between maths and other areas of the curriculum. This should enable you to deliver and display relevant and exciting lessons, motivating both yourself and children's appetite for maths.

Margaret Share

Snakes and Ladders

Focus of Learning

- Using number bonds and solving 'missing number' problems

Art and Display

1. Make a chequered background in the shape of a snakes and ladders board. Make three cardboard snakes and add an articulated snake's head, with wobbly eyes and forked tongue. Label the snakes in multiples of two, five and ten. Create ladders using art straws, and label the ladders 0–20. Add bead strings, arranged in groups of two.

2. Add a moveable sleeve decorated with a lizard that can slide up and down the snakes and ladders, hiding the multiples and digits. Give children a number of question cards with problems such as, 'How many twos make 10?' and '? – 3 = 16'. They should use the number lines and multiples on the snakes and ladders to help them.

Starting Points

- Using the display and question cards, ask children to write the missing numbers and other answers on individual whiteboards. Explain how to calculate the missing numbers using the display to support them. Add more than one sleeve to the snakes and ladders as children gain more confidence.
- Prepare a set of cards 1–10 and put them in a container decorated with snakes. Get children to individually pull out a card and ask the class to write the number needed to make 10 on their whiteboards. Ask children what other facts they know, particularly inverses: for example, 6 + 4 = 10; 10 – 4 = 6. To extend children's understanding, use number bonds to 20, 50 and 100.

Further Activities

- To support children in developing an understanding of number bonds to 10 and 20, put 20 pegs on a coat hanger. Show children the pegs and then display the problem 8 + ? = 20. Count eight pegs and leave a gap, then ask children to answer the problem by counting the remaining pegs to 12. Then show children the inverse operation 12 + 8 = 20. It can be useful to attach topic-related Clip Art or objects to the pegs. Alternatively, use bead strings instead of pegs.
- Provide slidey boxes (as shown) to support children in finding the missing numbers: for example, ? + 8 = 100 or 16 + ? = 20. Ensure you change the position of the missing box regularly. Also change the missing number for a missing symbol (for example, * or ? or a box) so that children understand that there can be a variety of signs used in this type of question.
- Laminate cards with empty boxes for children to write their own questions. Also ask children to write their own problems to illustrate their calculation. For example, 'There were 100 children playing in the playground. 70 went into school for their dinner. How many were left?' Children's questions and answers could be displayed in a problem solving book.
- Produce two sets of cards: set A has numbers (for example, 5–15); set B has the function (or operation) with a digit written on it (for example, + 6). Prepare a function machine, as displayed. Working in pairs, Child A picks a number card from set A (5–15) and fills in the 'In' section of the function machine (in this case, 3). Child B chooses a card from set B: the function (operation) with a digit (+ 6). Keeping this card secret, they fill in the 'Out' box of the function machine. Child A has to guess what the function was. They can check the answer using a calculator. For more able children, use larger numbers, negative numbers or decimals, and two or more functions.

IN	OPERATION	OUT
3	? (+6)	9

- Make a 2 × 5 grid. On the grid write the number bonds to 10, 20 or 100 (whichever is most appropriate). Play *Number Bond Bingo*. As you call out the appropriate number bonds, children can use counters to cover the appropriate number-matching bond. For example, if you call out 17, a child should cover 3 on their bingo card to make a number bond to 20. The first child to cover all the squares is the winner. Challenge children to tell you the inverse of their answer.
- Ask children to design, make and test their own *Snake Game*. They should include instructions and questions using numbers to 20, 50 or 100 in multiples of 2, 5 and 10. They can test the games and their instructions when they play with friends.

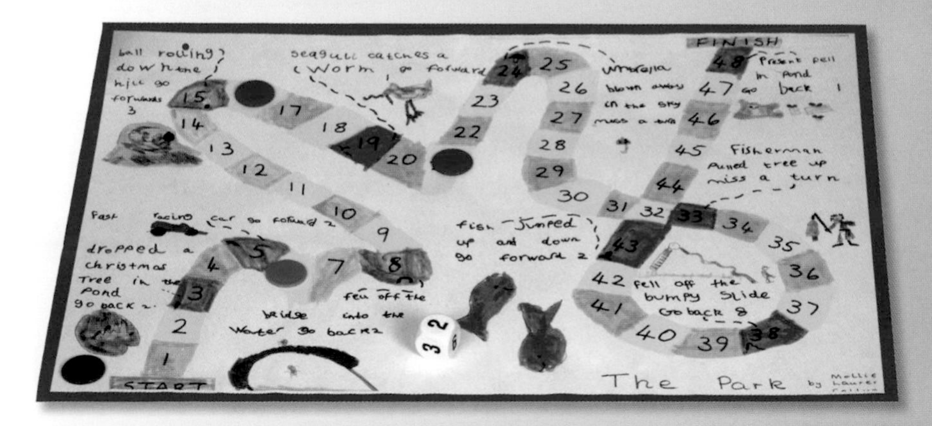

Working Wall

- Display a variety of multi-link cubes in two colours that illustrate some of the bonds to 10, 20 and 100. Give children laminated, cm² paper with 10, 20 and 100 squares (whichever is most appropriate for the child). Ask children to cover or colour a designated number of squares to show the bonds to 10, 20 or 100. Display some of the children's work on the wall or a table for reference.

Number Lines

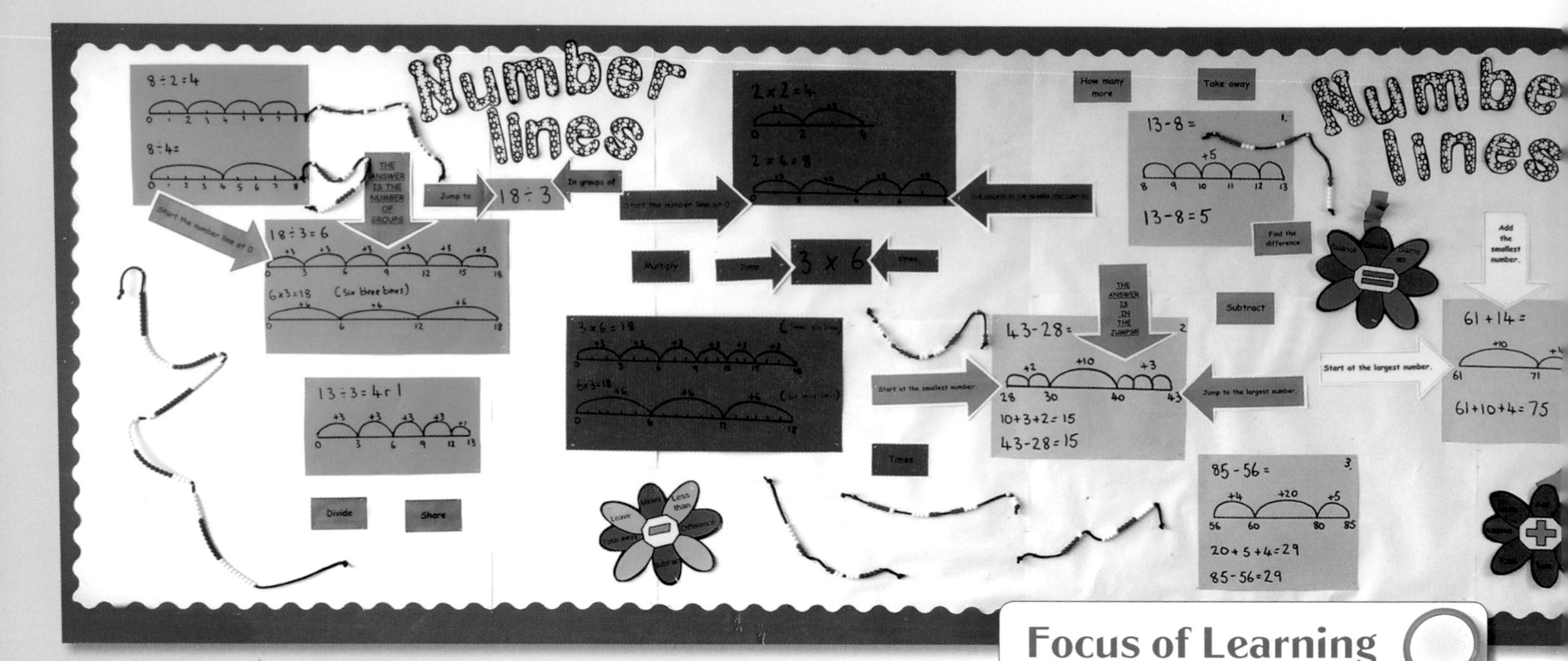

Focus of Learning

- Using number lines for addition, subtraction, multiplication and division
- Applying number lines in a history topic: for example, the Vikings

Art and Display

1. Create a display that will be relevant for an extended period of time. Produce examples of number lines and ensure they meet your school's preferred methods. Add vocabulary and tips to the display to support children in understanding any challenging parts of a concept.
2. Add bead strings that illustrate the relevant number line. Use these examples of number lines as models.
3. The number lines can be printed onto sticky labels and stuck into children's books, as examples for them to use as prompts when completing the calculation. You can also send the number lines home, to support parents in using the correct methods when helping their children with homework.

Starting Points

- Practise instant recall of number bonds to 10. Explain to children that you are going to hold up three fingers in the air. When you say 'Show me', they should answer how many more they will need to make 10: 3 + 7 = 10. This can be extended by explaining that each finger has a value of 10. Show four fingers and ask how many more they will need to make 100: for example, 40 + 60 = 100.
- Play *Number Bonds Bingo*. Ask children to choose six numbers between 1 and 9 and to write them on their whiteboards. Call out random numbers up to 10. If they have chosen a number to make 10, they cross it off. Write the number on the board for the class to check as the numbers are called. The first to cross off all the numbers is the winner.
- A variation is to give children a bingo card with eight numbers and then call out the number bonds. Checking is the same as the above game.

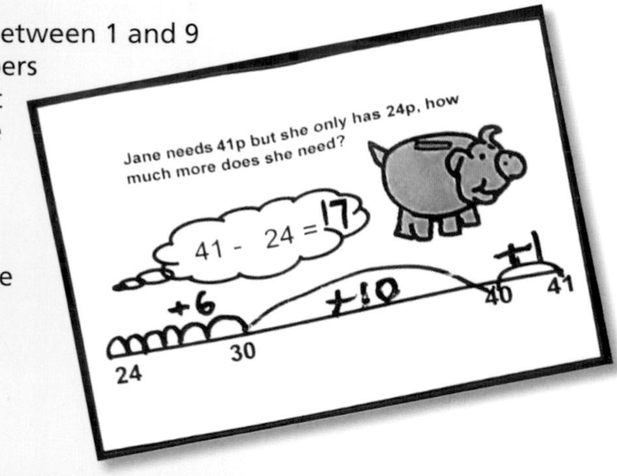

Further Activities

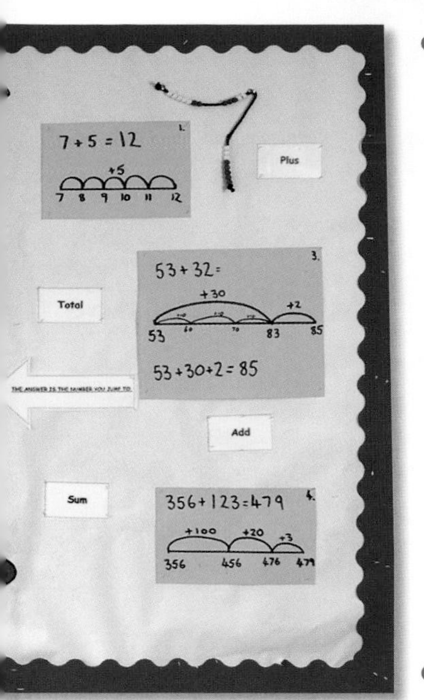

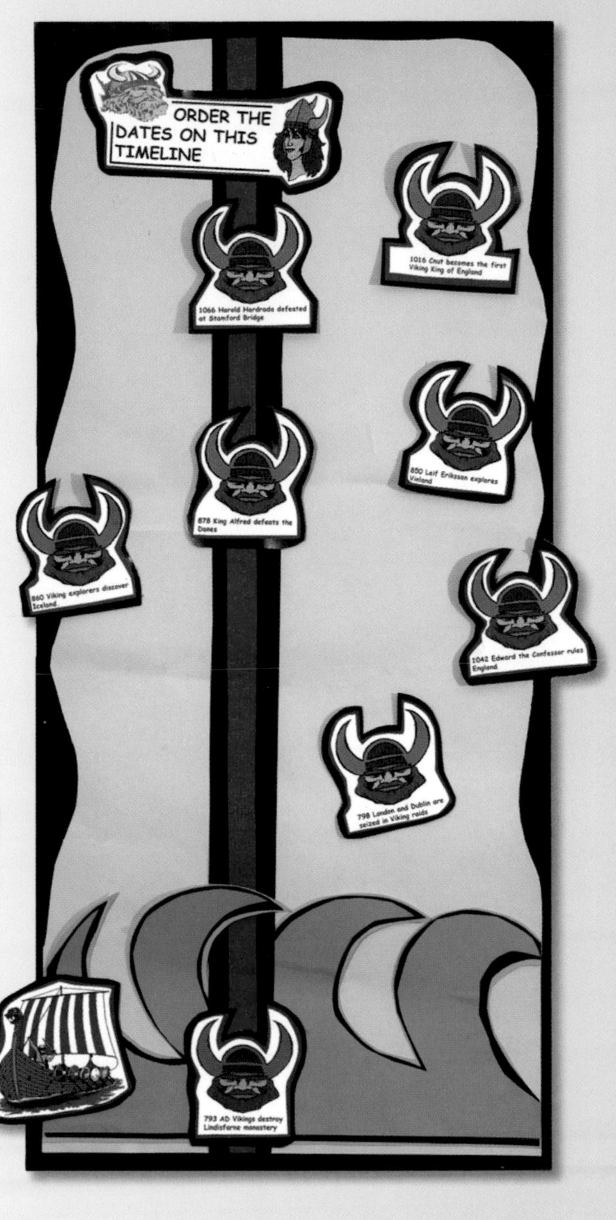

- To test subtraction by finding the difference and counting up, pose the following problem: 'Sue has 43 stickers and James has 28. How many more stickers does Sue have than James?' Ask children to count to 43 on their bead strings, marking the number with a peg. Then ask them to mark to 27 on their bead strings with another peg. Ask children to count the difference between the two numbers of beads: 43 – 27 is 16. Then show the calculation on a number line. Point out the challenging elements: explain that they should always start with the small number, put it in their head and then count up to the big number. The answer will be in the jumps at the top of the number line.
- Pose the following problem: 'Amy had 85 conkers and James had 48. How many more conkers does Amy have than James?' Invite two children to sit on chairs at the front. Give them cards of 82 and 48, and a washing line. Ask the class probing questions to check understanding. For example, 'Where will the child with 48 sit? Why? Where will the child with 85 sit? Why?' Ask children to hold the rope, which is the 'gap' or difference between the numbers. Then, using the rope as a number line, ask children to peg up the jumps to find the answer.
- Give children blank number lines, a variety of bead strings and laminated problem cards. Provide cards at the appropriate level and fill in part of the calculation on some cards for children to complete. Depending on their ability, children can check the answer using the inverse or a calculator, or they can show you. Display examples of this work.

Working Wall

- Using Clip Art, produce images of a Viking with dates of important events. Children can order the Viking dates on a pre-prepared timeline using Blu Tack. Ask questions relating to time differences between events. For example, 'How long after King Alfred defeated the Danes did Edward the Confessor rule England?' More dates can be added as the project progresses.

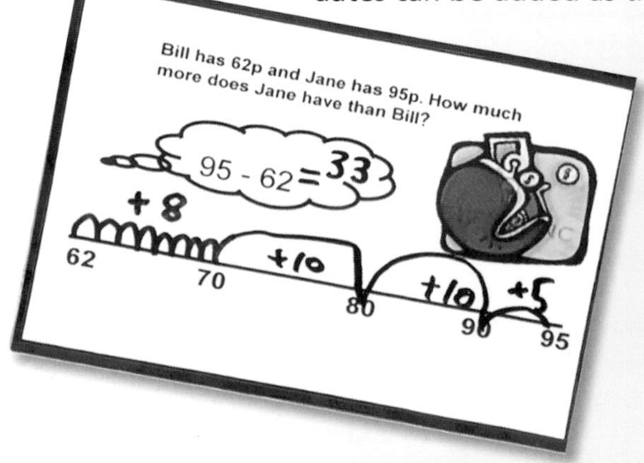

Cross-curricular Links

- **HISTORY** – Use the concept of finding the difference by counting on to calculate time differences in history. Add a timeline, then add questions relating to the events in the period of history. Provide laminated empty number lines for children to use to calculate the difference. You can also present number lines in different forms: for example, on crowns as a Viking king's timeline.

Maths Facts Wizard

Art and Display

1. Create a castle shape with windows. Design the windows as a jigsaw puzzle as shown. Add a wizard and stars to the display.
2. Put cards in the windows to demonstrate the inverses of the appropriate maths facts. Add problem cards for children to solve.

Focus of Learning

- Developing instant recall of multiplication, division, addition and subtraction facts, and checking using inverse operations

Starting Points

- Give children a collection of small bags with three sweets in each bag (if children are learning the three times table). Recording their answers on individual whiteboards, ask children to solve problems such as, 'How many sweets will I have in three bags?' Demonstrate how to do this by taking the sweets out, keeping them in threes and counting 3, 6, 9, etc. Ask further questions such as, 'How many sweets would I have in six bags? I have 24 sweets. How many bags of three sweets could I make?' Count in threes and model on a number line.
- Give children a collection of bags weighing, for example, 7g (for children learning the seven times table). Ask questions such as, 'How many bags would weigh 21g? How many grams in five bags?'
- Display a collection of multi-link cubes arranged into different columns. For example, six columns with eight multi-link cubes in each column. Ask children to write down a calculation for each multi-link model and then create a problem to fit the calculation. For example, 'John has six packets of stickers with eight stickers in each packet. How many stickers has he got altogether?' Display the problem with the appropriate multi-link cubes.

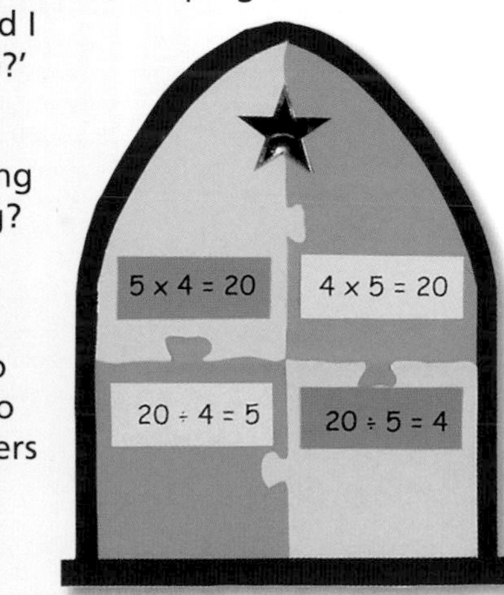

Further Activities

- Produce a calculation such as 4 + 20 = 24. Add a wizard with stars as a sleeve that moves up and down the card to hide the numbers. Children then calculate the hidden numbers.
- Play *Wizards and Dragons*. Divide the class into two teams and ask one child to represent each team. Write a selection of answers to multiplication and division times tables on the board. Children should stand at opposite sides of the board, ready to draw a ring around the appropriate answer to the question. Read out a question such as, 'How many sevens in 21?' The first child to circle the answer (3) is the winner. Swap children around for each question.
- Draw a 6 × 2 grid, approximately 40cm by 40cm, on the playground. Add the multiples of eight from 1 to 12 in each square. Prepare a set of question cards relating to the eight times table. Ask a child to read out a question as another child picks up a beanbag and puts it in the correct place on the grid. An alternative version of the game is to ask children to write the answer on a whiteboard and to stand at the appropriate position on the grid with their whiteboard. At the end of the game, children holding the whiteboards on the grid say the table forwards and backwards. Give an extra point to children who use their multiplication or division fact in a problem. For example, 6 × 8 could be, 'John counted 48 spiders' legs altogether. How many spiders would that be?'
- An easy-to-organise game is to ask children to write down five facts from a specific times table on individual whiteboards. Ask children questions such as, 'What's 5 × 3?' If they have the answer 15 on their whiteboard, they should strike it through. The first child to finish striking through all of their number facts is the winner.
- Prepare a set of cards with multiplication or division problems at the appropriate level. For example, 'James has 64 marbles. He divided the marbles equally between eight people, including himself. How many marbles did each person have?' Also, prepare cards showing the symbols for multiplication and division. Ask children to match a problem to a symbol card. They should justify their reasoning using a 'because' card that supports them in structuring their explanation. A challenge could be for children to complete a couple of the questions and check them using the inverse. Children who finish early might also enjoy devising questions for other children to categorise: for example, +, ×, ÷, +.

Working Wall

- Create a section of the castle wall as a working wall and provide laminated, computer-generated empty box questions for children to fill in. For example:

?	+	*	= 64
36	÷	?	= 6
96	–	?	= 50

- Working in pairs, one child completes all the questions on the wall, and the other child times the activity and checks the answers. Ask children to write their own empty box questions for others to fill in.

Gingerbread Arrays

Focus of Learning

- Using arrays to support children in understanding the relationship between multiplication and division
- Using arrays to support worded problem-solving

Art and Display

1. Make and varnish 20 gingerbread men. Alternatively, cut and laminate cardboard gingerbread men.
2. Cut out a gingerbread-man freeze and attach it to the display at the top and bottom.
3. Make a variety of work cards, as shown, with worded problems for children to solve. For example, display the recipe for six gingerbread men and ask children to solve problems relating to the recipe by doubling and halving.
4. Give children small bags of gingerbread men to use as arrays to help them to solve the worded problems.

Starting Points

- Show children an array of objects from 5 × 5 to 10 × 10. The objects could be images of gingerbread men taken from the internet. Use different images according to the class topic and the age of children: for example, dinosaurs or footballs. Laminate the arrays and use them to direct the children. For example, 'Show an array of 6 × 5 gingerbread men.' Then ask children to write down the multiplication fact. Clearly demonstrate the difference between arrays showing 6 × 5 and 5 × 6, stressing that 5 × 6 is 'five, six times' . Demonstrate this difference by using the gingerbread men from the display and putting them into separate bags.
- Give groups of four children multi-link cubes. As a group, invite them to show the following calculations using the cubes: 4 × 5; 3 × 6; 6 × 7. For the first calculation, children can label the array 4 × 5, saying '4 multiplied by 5' . They then write a suitable problem for their array. Conversely, give children multi-link cube arrays and ask them to write the calculation and then make their own problem for other children to solve. In both

activities, it is important that children draw the array as well as writing the problem.

Further Activities

- Give children a variety of arrays and invite them to make up their own worded problems. The challenge could be to devise problems that need two or more steps to solve. You could display the children's problems around the arrays on the working wall.
- Children's worded problems could also be gathered together and displayed in a class book of multiplication problems, showing the array, calculation and number of steps. Answers could be hidden behind a flap.
- Ask children to make different arrays using pegboards and pegs or to use the gingerbread men from the display. They could then make up a problem using their array. For example, 'Thirty coloured pencils need to be shared equally between six children. How many coloured pencils for each child?'
- Play *Catch the Gingerbread Man*. Invite a group of children to stand in four rows with three children in each row. Ask one child to be the gingerbread man and one to catch him. Start the game with the hunter at one corner of the group, and the gingerbread man at the opposite corner. The group will hold hands in a row with three people. When you say 'Go' the gingerbread man runs down the corridors made by children holding hands, with the hunter chasing him. When you call out 'Change' , children drop their hands, turn to the side, and hold hands in rows of four, thus stopping the hunter. During the game, ask children to 'Freeze' and ask questions such as, 'What array are we making now?' They should answer 4 × 3, which is four children in three rows, or 3 × 4, which is three children in four rows. Repeat this activity with different arrays of children up to 24: that is, rows of 6 × 4. Point out the division facts in the array 24 ÷ 6 = 4.

Working Wall

- Create a working wall that shows very clear models for multiplying, dividing and using appropriate mathematical vocabulary. Children's whiteboard work illustrates arrays. This can then be clarified by using multi-link cubes, which show the calculation 3 × 4; three multiplied by four; three, four times. You can display a section of your school's calculation policy: in this example, the working wall was used to illustrate repeated addition and appropriate vocabulary.

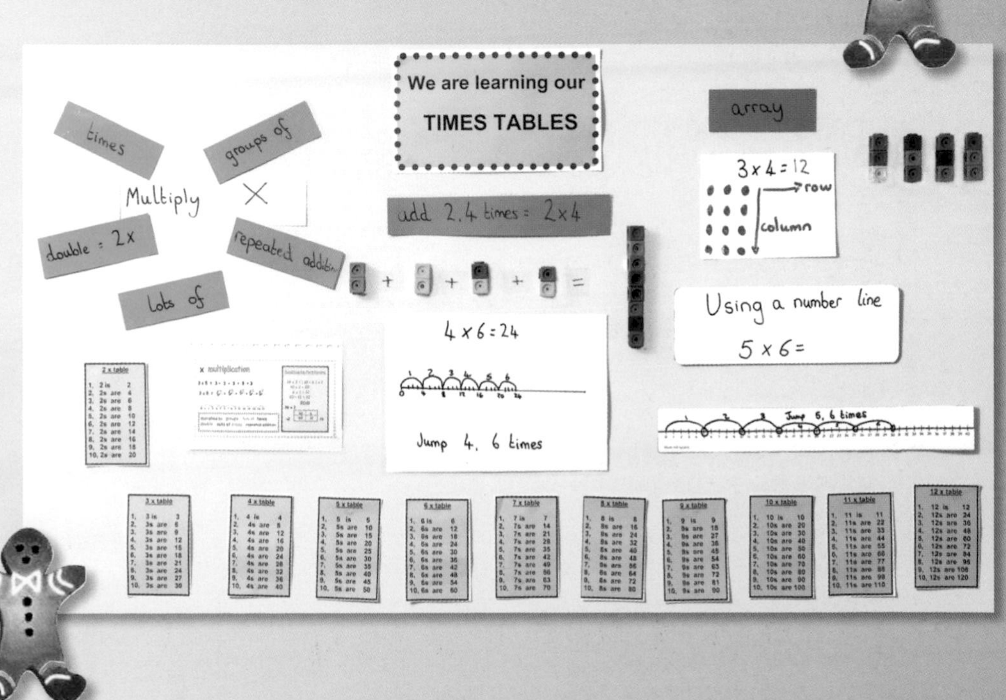

Terrific Tables

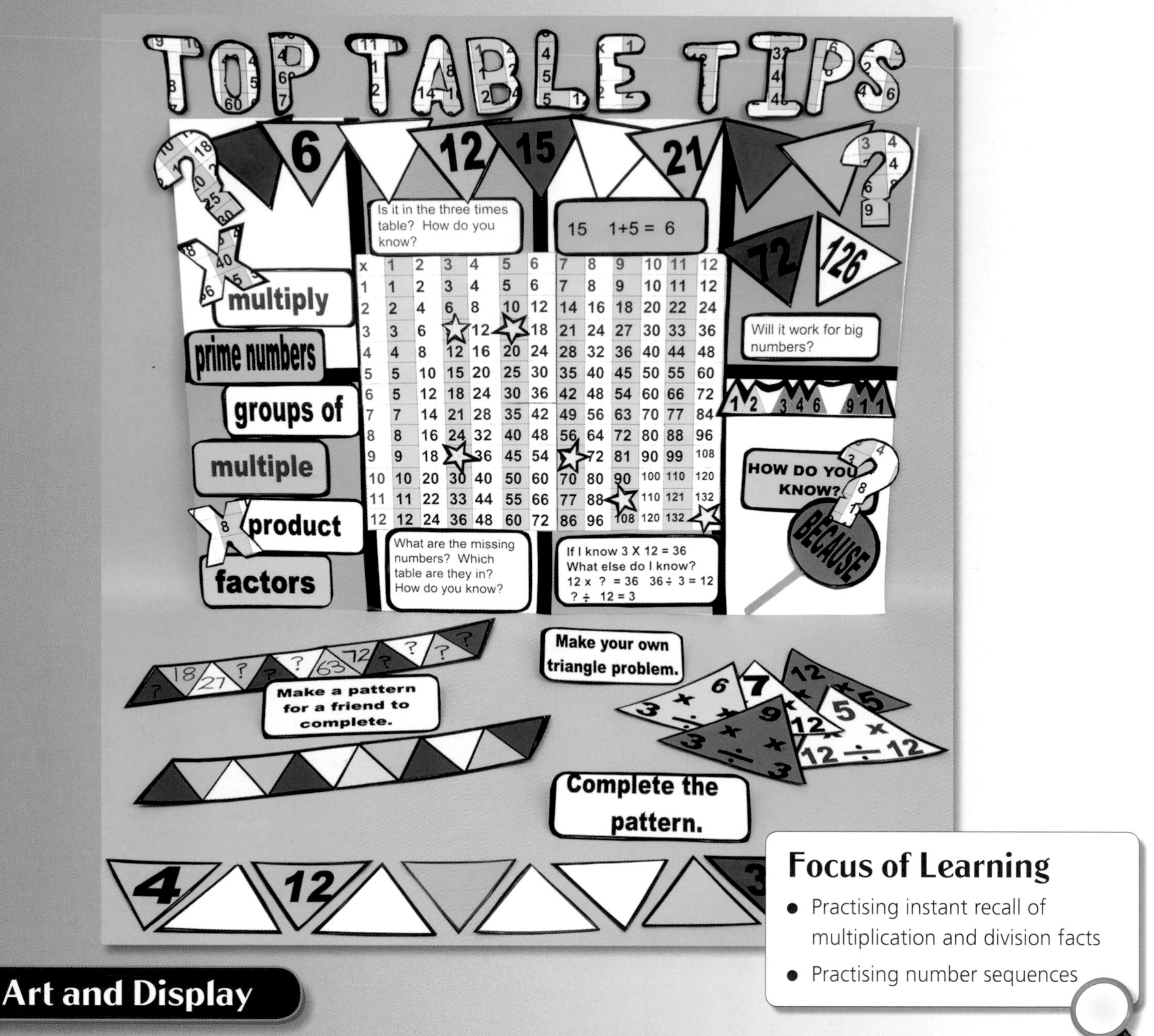

Focus of Learning

- Practising instant recall of multiplication and division facts
- Practising number sequences

Art and Display

1. Produce a Top Table Tips display. Add a multiplication grid and appropriate vocabulary. Attach with Blu Tack so the sequence can be changed on a regular basis. To find out if a number is in the three times table, add up the digits of the number: for example, for 15, 1 + 5 = 6. If the numbers add up to 3, 6 or 9, then you know the number is in the three times table. Ask challenging questions: for example, 'Is 12 346 911 in the three times table? How do you know?'
2. Add number sequence questions and laminate for children to fill in. Prepare triangular multiplication sequences for children to complete, and encourage them to devise their own questions.

Starting Points

- Using a counting stick, put one finger on one end of the stick, saying, 'If this end represents 0, and the opposite end represents 30, which multiples am I counting in?' Place your finger over the middle of the counting stick and ask, 'What's special about this number? (It's the halfway number.) Because I know the halfway number 3 × 5 (three, five times), what other multiples do I know?' (3 × 6, which is three more, and 3 × 4, which is three less than the halfway number.) Point out that if 3 × 10 is 30, 3 × 9 is three less than 30, so is 27, which is easier to remember. Repeat this type of activity for all the tables.

Further Activities

- Write a number sequence on the whiteboard or board: for example, *, 6, *, 12, 15, * , 21, * ,*. Ask children what the next term in the sequence is and how they know. Explain that a sequence is a set of numbers arranged in order according to a rule. Each of the numbers in the sequence is called a term. The rule here is +3. The next term in the sequence will be 30, which is 3 × 10. Encourage children to use the correct terminology.
- Children can play the game *Helping Hands*. They sit in pairs opposite each other with a third child checking the answers. On the command 'Show me' , both children show fingers to represent half of the multiplication table. For example, for 6 × 5, one child will show six fingers, the other five. The first of the two children to whisper the answer is the winner. The third child checks the answer using a calculator.

Working Wall

- Prepare a set of eight 100 squares as a working wall display. Illustrate the number patterns for 2, 3, 4, 5, 6, 7, 8 and 9. Ask questions about specific tables: for example, 'Is 99 a multiple of 9?' Children use the 'because' card to structure their answer. For example, 'I know that 99 is a multiple of 9 because 9 + 9 = 18, which is a multiple of 9' . Explain that there is no easy way to learn the seven times table, but that the challenging 7 × 8 = 56 can be remembered by knowing 5, 6, 7, 8.
- Invite children to investigate the patterns that are formed by adding or multiplying multiples of 2, 3, 4, 5, 6, 7, 8, 9, and the ways to support learning the multiples.

Spider Race (Probability)

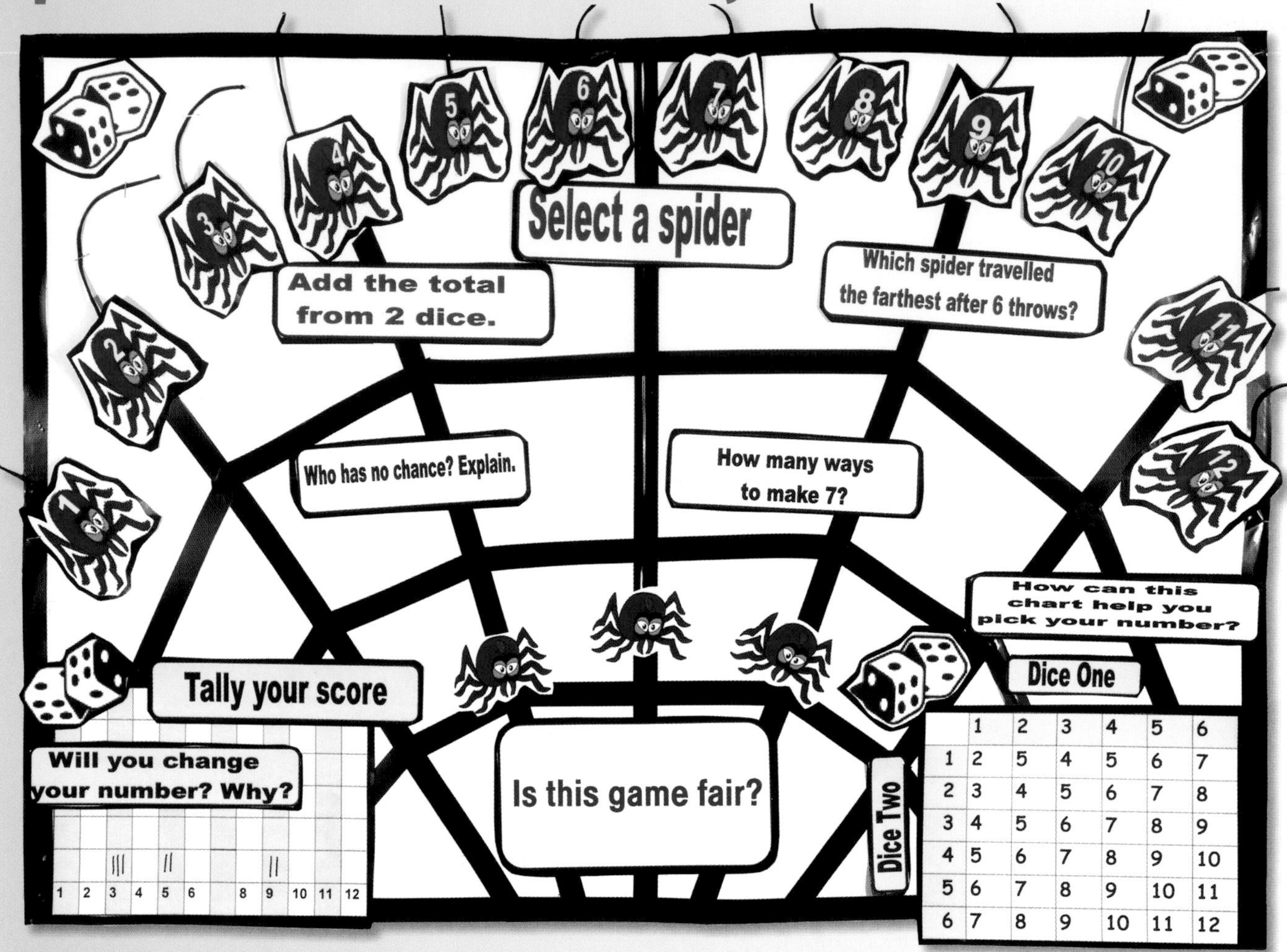

Art and Display

Focus of Learning

- Understanding probability and its application

1. Produce a spider's web game, as a starting point for children to make an informed decision about whether a game is fair or not. Each child chooses one of the spiders numbered 1–12. Two dice numbered 1–6 are thrown in turn and each child adds up the total on the dice. If the number on their spider is thrown, they can make a tally. For example, 4 + 5 = 9, so they will write a tally on the 9 column, which represents the number 9 spider. The spider with the most tallies after a given number of throws (for example, 12) is the winner.
2. You could also use a 12 × 12 grid as a baseboard for the game, and the children move their spider one square each time their number is thrown. A child who has chosen 1 won't have a chance because the smallest number that can be scored with the two dice is 2. Children can investigate the probability of each number being thrown by investigating the combinations of numbers thrown to make a specific number: for example, 9 = 6 + 3 or 5 + 4. Children make a findings board that can be used to choose a spider, or to set up a system to make the game fairer: for example, choosing to leave the number 1 out of the game.
3. Produce a display to illustrate the game. Use electricians' black tape to make a spider's web.

Starting Points

- Give children a coin. Ask them to work out what the probability is of them rolling heads after 10 rolls. The probability is ½ – or the number of ways that it can happen over the number of outcomes. Ask children to carry out a variety of experiments:

- Rolling a single dice, ask them, 'What's the probability of rolling a 3?' Make an eight-sided spinner numbered 1–8 and ask, 'What's the probability of an 8? Why?'
- Give a group of children a selection of lollipop sticks: six red, three black, one green and one blue. Put them in a bag and ask children to guess the probability of a red, black, green or blue lollipop stick being chosen first. Ask them to justify their answers.

Further Activities

- In groups of 12, give children two dice, spiders labelled 1–12 and an empty tally chart. Children roll the dice, calculate the total and, each time the dice shows their spider's number, write it on the tally chart. For example, if 4 + 5 is thrown, they mark one tally on their chart against 9, 4 and 5. The winner is the spider that has the most tallies after ten rolls of the dice. Ask children questions such as, 'Which spiders have the most chance of winning? Why? Which spiders have a poor chance of winning? Why?' Ask children to explain to a friend why 9 would be a good number to choose, and why 1 wouldn't be a good choice. Challenge children to think of ways to make the game fair. They may have ideas such as not using number 1 or giving some spiders a headstart in the race. This game can be used in a variety of contexts: for example, car racing, horse racing or flat racing.
- Play the strategy game *Probability – How to Win* as a table-top activity. Make a 4 × 3 grid, using electricians' tape to make the compartments. Make a set of 1–12 cards and give children dice and cubes to play the game. Children will soon realise they need to place the numbers that have the most combinations close together: for example, 8, 9 and 6.

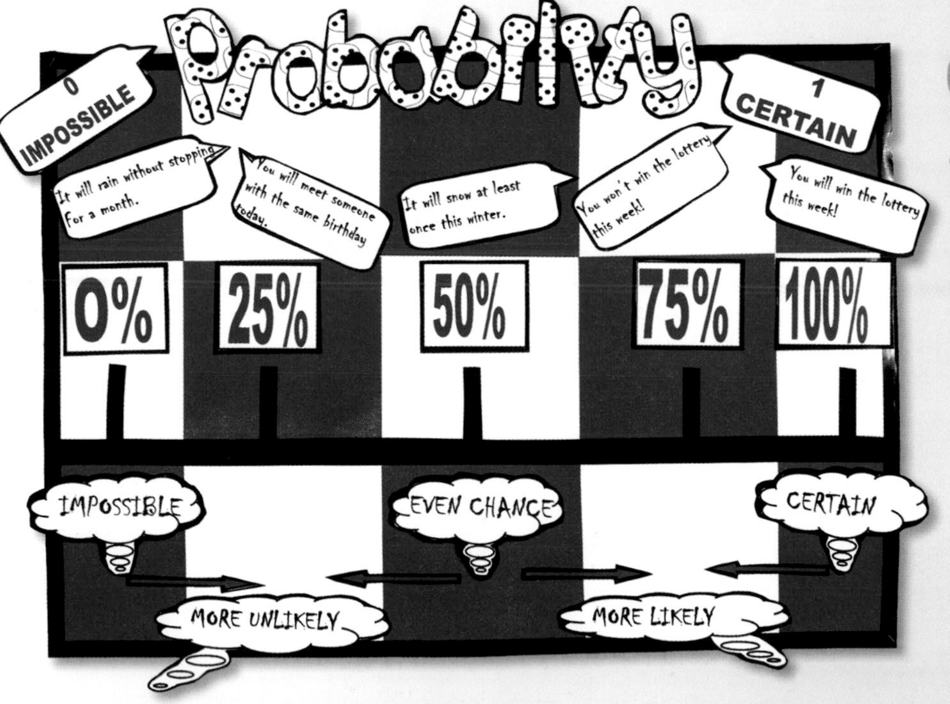

Working Wall

- Give children a set of questions to place on the probability line as shown (see instructions on image). Then, working in pairs, ask children to devise their own set of probability questions for other children to place on the probability line. Provide small coloured card bags to hold the sets of questions. Ask children to answer questions using the language of probability. For example, 'What's the probability of it raining tomorrow? What's the likelihood of seeing a car on your way home? What's the likelihood of meeting the queen?'

Cross-curricular Links

- **PSHCE** – Play probability games using dice rolling and coin tossing.
- **LITERACY** – Write instructions for playing games.

Galaxy Travel

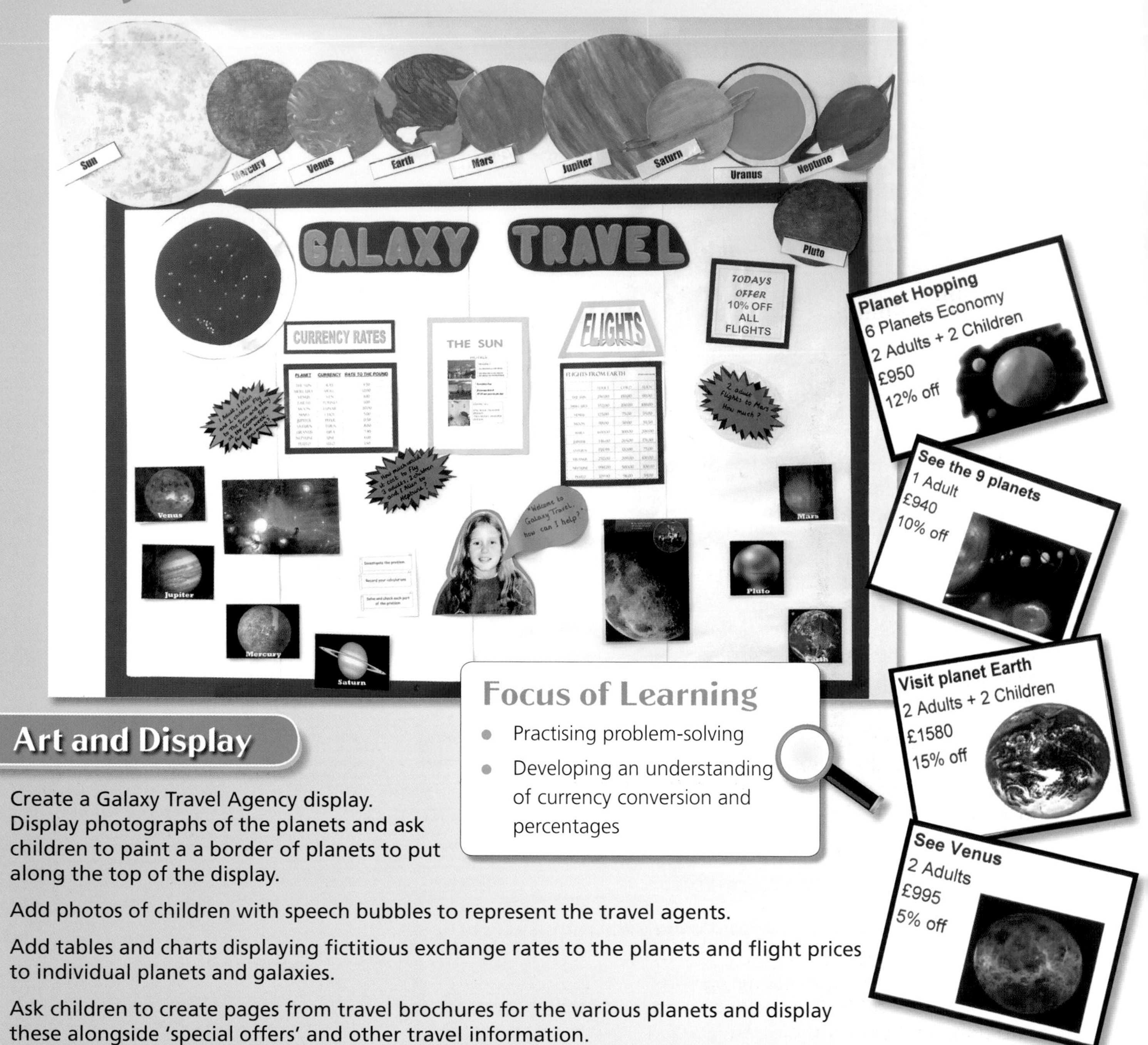

Focus of Learning

- Practising problem-solving
- Developing an understanding of currency conversion and percentages

Art and Display

1. Create a Galaxy Travel Agency display. Display photographs of the planets and ask children to paint a a border of planets to put along the top of the display.
2. Add photos of children with speech bubbles to represent the travel agents.
3. Add tables and charts displaying fictitious exchange rates to the planets and flight prices to individual planets and galaxies.
4. Ask children to create pages from travel brochures for the various planets and display these alongside 'special offers' and other travel information.

Starting Points

- Use the display as a starting point for role-play activities relating to choosing the most appropriate holiday, working out the cost of a holiday and finding the difference in prices of holidays. For example, ask, 'What's the difference in price of a visit to Earth at £1580 with 15% off? How much does it cost to go to Venus at £995 with 5% off?'
- Use percentages in a real-life situation. For example, give children up-to-date travel brochures and ask them to price holidays for two adults and two children for a week in Florida.
- More able children can work with the fictitious exchange rates on the display. For example, ask, 'How much would it cost in *Une* for two adults to visit the Cosmic Spa hotel on Neptune?' Less able children can plan holidays using the pound as currency. For example, 'How much would it cost for two adults to visit the Cosmic Spa hotel for two weeks?'

Further Activities

- Give pairs of children small, blank, laminated 100 squares and invite them to fill in a selection of percentages: 10%, 25%, 35%, 12%, 1%, etc. Alternatively, provide small laminated squares with the above percentages filled in for them to identify. Draw attention to the fraction, decimal and percentage equivalents.
- Give children dice marked 1%, 5%, 10%, 20%,15%, 25% and a blank 100 square. Children take turns to roll the dice, then ask them to fill in and name the percentage shown on the dice. The first player to fill in the grid is the winner. For a further challenge, provide a grid with 50 squares.
- Choose groups of up to four children to work in the Galaxy Travel Agency (it works best if the 'travel agent' is more able than the 'customers'). The customers should plan, choose and price holidays, and calculate the prices with percentage discounts. Travel agents should use a problem solving frame that will support their customers in working out their calculation.

Working Wall

- Prepare a working wall to support children in calculating a variety of percentages from 50% to 1%. Provide examples of holidays with specific percentages to take off. Give children 100 squares so that they can work out specific percentages, colouring in and writing the percentage. For example: 2% and 18%, and percentages that then total 100%. These can be arranged around the display. Encourage children to answer all the questions, and display their answers on the wall.

Cross-curricular Links

- SCIENCE – Prepare a booklet detailing the properties of the different planets. Children can use it when giving advice to customers in the travel agency.
- ART – Paint the different planets and show their features.
- LITERACY – Research the different planets so children can discuss and advise when role-playing at the Galaxy Travel Agency. They could also create travel guides to the planets.

Party Hats

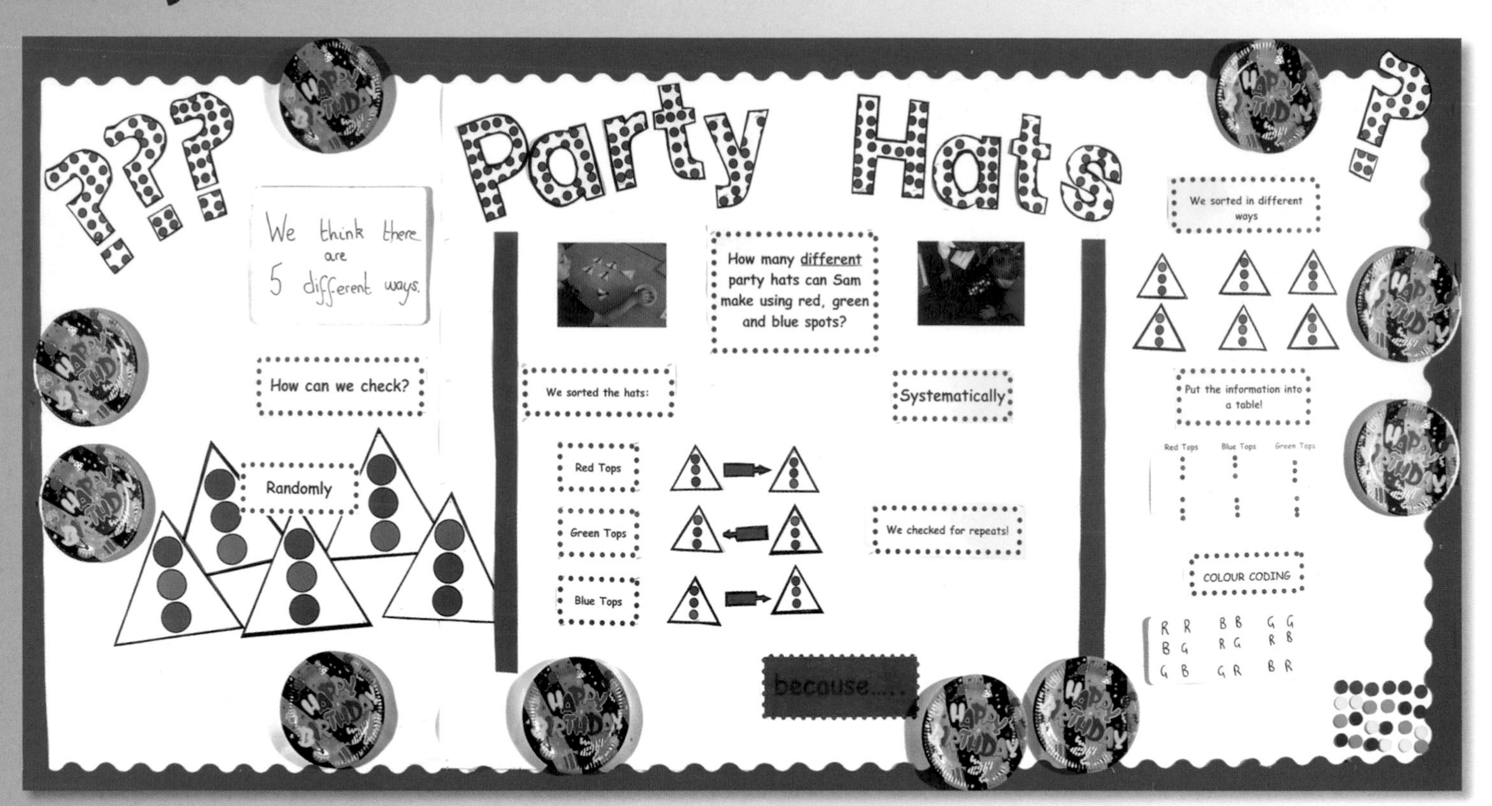

Art and Display

1. Ask children, 'How many different party hats can we make using three colours of bobbles, using each colour only once?' Limiting the number is a structured way of introducing sorting and classifying. When children are more confident in this, give them the three colours without restrictions. Prepare sets of small, laminated party hats with combinations of one red, blue and green bobble.
2. Ask children to sort the bobbles and see if they can find all the colour combinations of them.
3. Using A3 paper, ask children to draw their combinations in a systematic way. Invite them to explain their work and add their presentation to the display. Children should also add labels explaining how they checked and sorted their work.

Focus of Learning

- Sorting and classifying systematically
- Drawing diagrams and tables to solve problems

Starting Points

- For less able children, small groups could use the hall or an outside space for the following activity. Ask those with the red tops to stand at the front, then children with blue tops, then green tops. Take photographs of these groups and display next to children's written explanations of what they have seen.

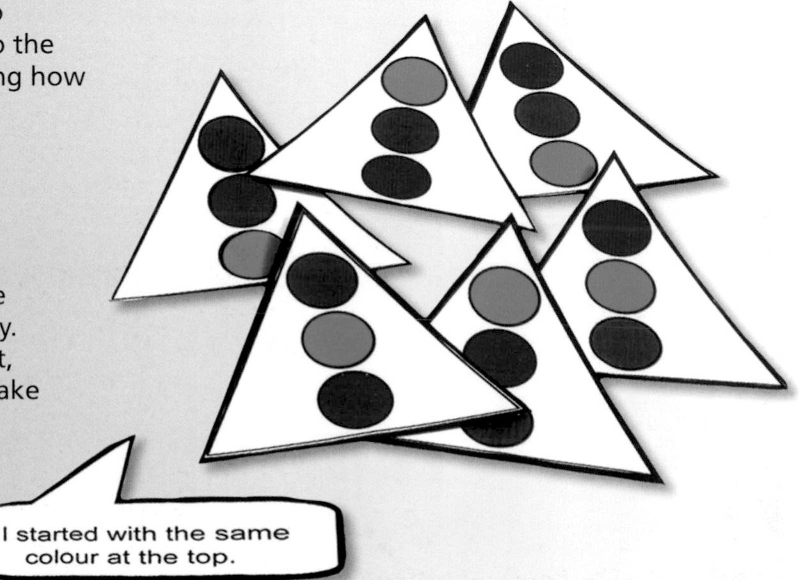

Further Activities

- Invite two children to come to the front and hold a card with a 9 or a 6 on it. Ask the class, 'How many different numbers can we make using nine and six?' Children can write down the possible numbers on individual whiteboards. Ask them to explain how they know there will be two numbers. Next, ask three children to come to the front and hold up a 9, 6 or 5. Ask children to make as many numbers as they can using nine, six or five, writing the numbers on their whiteboards. Ask them to show and explain their calculation. Look for children who are starting to show a systematic strategy and who can explain this. Ask, 'How do you know that you have found all the combinations of numbers?' Then challenge them to consider how many different possible numbers there are using four digits.
- Play *Place Value Finder*. Give children calculators and explain that you are going to enter a three-digit secret number into your calculator: in this example, 365. Children discover the secret number by guessing, using place value, then entering the number into their calculators. They can ask you questions such as, 'Is there a 3 in your number?' With 365, the answer would be 'There is a 3: 300' . So 300 is deducted from your calculator and added to the children's calculator, continuing in the same way for 6, 10 and 5 units. The aim is to get to the number in no more than five guesses.

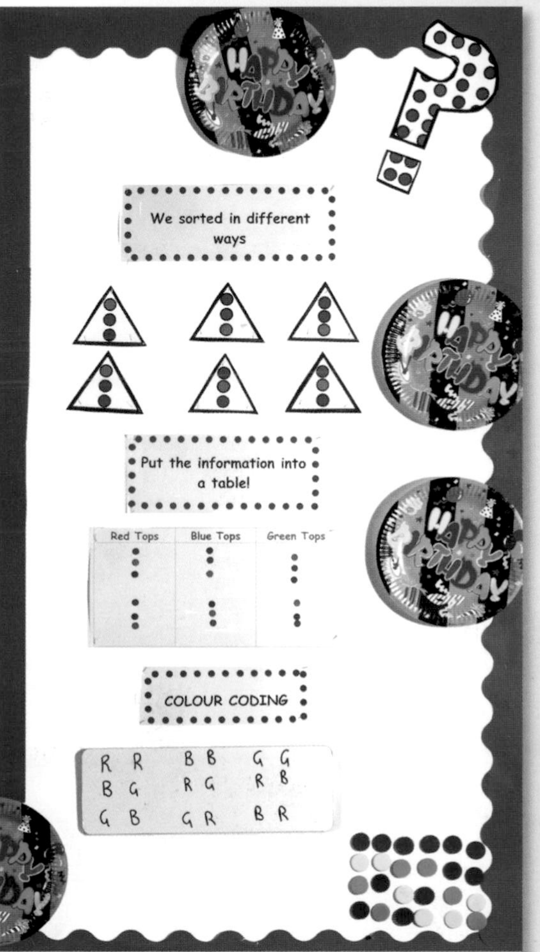

- Give children word problems that involve being systematic in finding the solution. For example, 'Sam has been invited to a birthday party but the street number has been torn into three different digits. How many houses will he have to visit before he finds the correct address?' Ask children to show their working out and explain their strategies. Another example is, 'John's mum bought some small buns for a party. There were square chocolate cakes and triangular cherry cakes. She knew that there were 24 sides in her box. How many of the cakes could be chocolate cakes and how many cherry cakes?' Give children laminated paper cakes to help them to solve the problem.
- Create a challenge table top as shown. Prepare a selection of red, green and blue counters, and some questions for children to answer. Children use the labels to help them to structure their systematic recording. Challenge more able children by asking, 'How many different hats could be made if we had four spots?' Encourage them to make a list of helpful material: for example: hats with four empty spots and four counters, and a table to record their findings.

Working Wall

- Pose the following problem: 'How many different party hats can we make using three different colours of bobbles, using each colour only once?' You can extend the problem by having four different-coloured bobbles. Ask children to explain their answers to the problems and then attach them to the working wall. Produce some labels to explain their thinking, such as: 'We sorted in different ways', 'We put the information in a table' and 'We colour coded'.

Dem Bones

Art and Display

1. Create a paper skeleton 88cm tall. The head should measure 11cm and the body 77cm. Label the skeleton on the display with these measurements. Prepare an extra skull measuring 11cm and two pieces of card 77cm long and 88cm long respectively, which the children can use on the floor to support measuring the proportion of the skull to the body.
2. Mathematically, one 'skull' in every 'body' would be $\frac{1}{8}$.

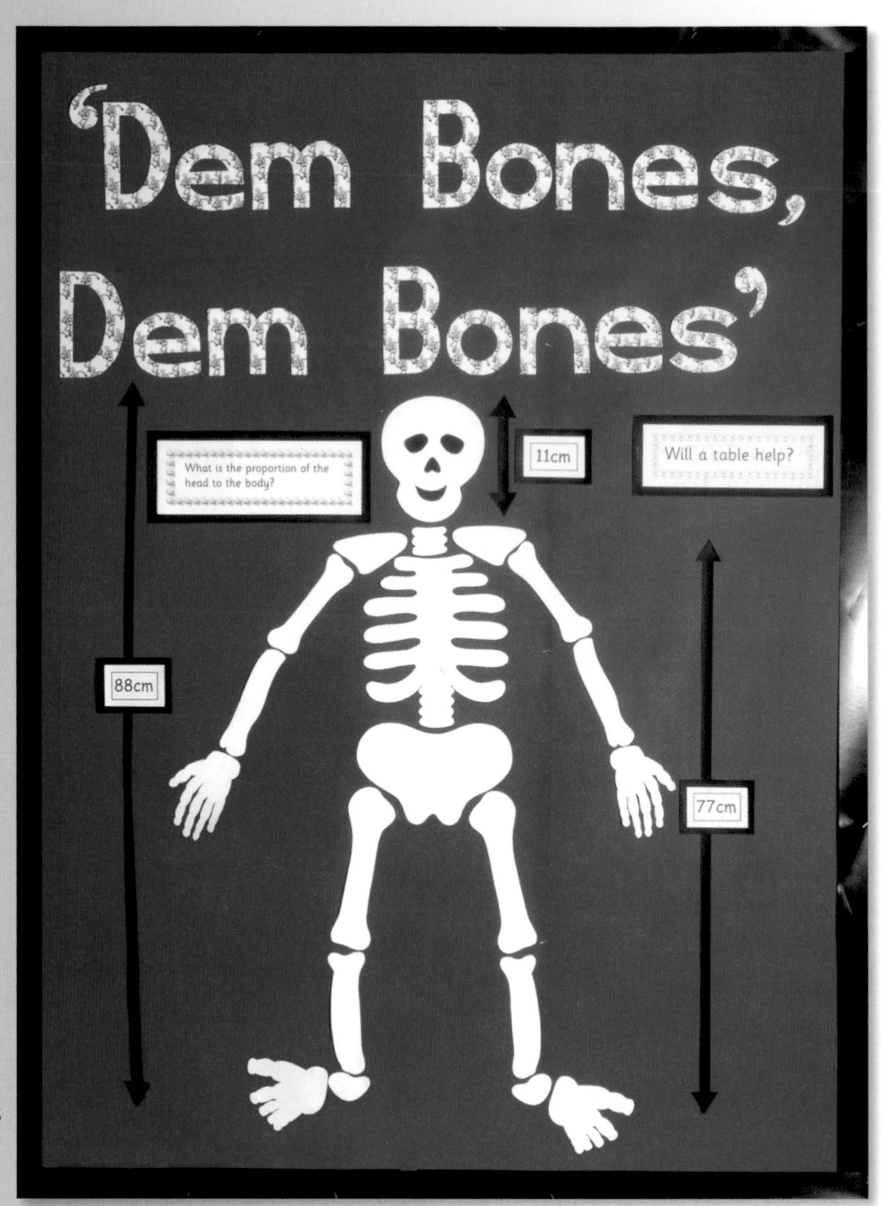

Starting Points

- Prepare ten bones in card, with two bones being rib cages. Explain to children that the proportion of the rib cages and skulls is two out of ten, because there were two rib cages and eight skulls: therefore $\frac{2}{10}$ or $\frac{1}{5}$ are rib cages. Generate different proportion questions using the bones. Remember to use the language of proportion, such as 'in every', 'for every', 'to every', and so on.

- Give children cubes so that they can generate their own proportion statements such as, 'The proportion of the black cubes is $\frac{2}{5}$ for every three white tiles: therefore, two black tiles to every three white tiles.' Provide interlocking cubes to demonstrate fractions. For example, arrange two sticks of seven (with two colours) to demonstrate the proportion that one in every seven cubes is green, so as a fraction $\frac{1}{7}$ is green. Ask children to make their own proportion sticks and write statements on individual whiteboards. For example, 'My stick shows that two in every ten cubes is yellow. As a fraction, it's written $\frac{2}{10}$ or $\frac{1}{5}$.'

- Invite children to draw around their hands. Cut out the hand and explain that one in every five digits is a thumb. Attach the hands to the main display and add questions such as, 'How many fingers will there be if I can see four thumbs?' This will enhance the display and support children's understanding of proportion.

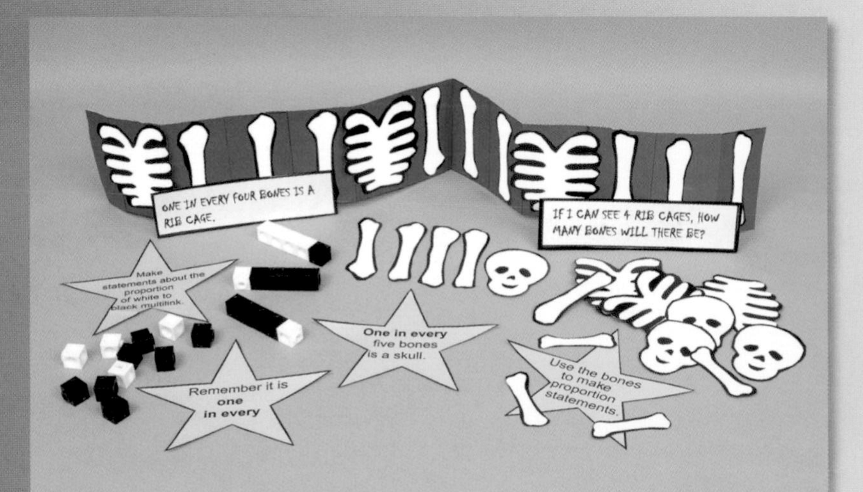

Focus of Learning

- Understanding ratio and proportion

Further Activities

- Ask children to calculate the height of a model skeleton. Provide a laminated table for them to fill in to support the calculation. Pose problems such as, 'I'm only 25cm, which is one-quarter of my size. How tall am I?'
- Demonstrate finding the proportion of the head to the body of a 30cm doll (or other suitable toy). Invite a child to measure the doll from head to foot, and then measure just the head. Ask children to record the measurements on whiteboards and ask children to calculate the measurement of the rest of the body. Then use paper strips of the head measurement to demonstrate the proportion of the head to the body. For example, if the head is 10cm and the whole doll is 30cm, the proportion is one in three.
- Put large sheets of paper on the floor and ask children to work in groups of five. One child should lie flat on the floor and the other children can draw around them. The child being measured stands up and the length of their head, then their body, is measured using a tape measure. The figures should be entered into a table: for example, head 22cm, height 154cm, rest of body 132cm. The proportion here is 22/154, or one in seven. More manageable investigations could be the proportion of the hand to the arm, measuring from the shoulder to the tip of the finger.
- Explain that ratio is very useful in everyday life; give the example of mixing drinks. Demonstrate this using orange squash. Make up a drink using one tablespoon of orange and eight tablespoons of water. Explain that the ratio in this case is 1:8: one part of cordial for every eight of water. Give children a variety of different cordials and encourage them to try different ratios of cordial to water to find the most pleasant-tasting drink. Children could then make a list of their favourite cordial mixes including their ratios, or they could make a table displaying the ratios.

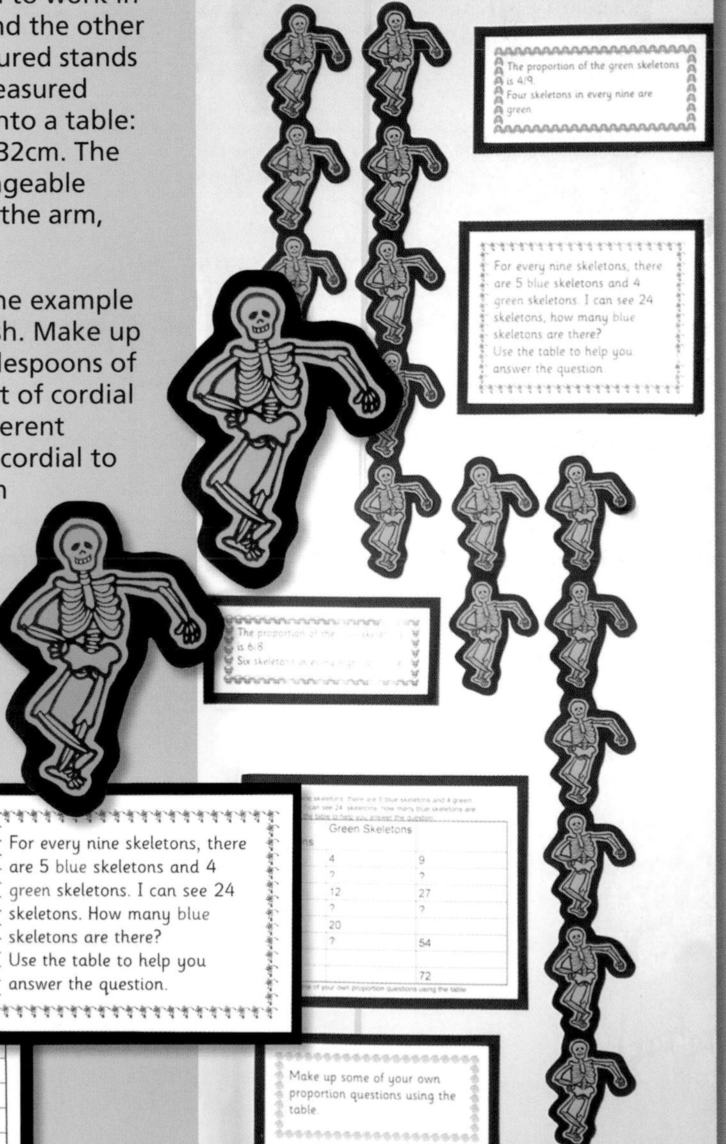

Working Wall

- Prepare two groups of skeletons: one representing four in every nine skeletons in green, and one representing six in every eight skeletons in blue. Prepare a table to support children in answering questions such as, 'If there were 20 blue skeletons, how many green skeletons would there be?'

For every nine skeletons, there are 5 blue skeletons and 4 green skeletons. I can see 24 skeletons. How many blue skeletons are there? Use the table to help you answer the question.

Blue Skeletons	Green Skeletons	
5	4	9
10	?	?
?	12	27
20	?	?
25	20	
30	?	54
?		
40		72

Make up some of your own proportion questions using the table.

Fantastic Function Machines

Art and Display

1. Prepare a Fantastic Function Machine display. Use circus animals for the function machine operations × 9 and + 12. If you use Blu Tack to add the functions, they can be changed on a daily basis if necessary. Prepare circus balls in a variety of colours to act as the start and answer numbers. Prepare questions as displayed.

Focus of Learning

- Applying inverse operations to multiplication, addition, division and subtraction

Starting Points

- Ask three children to sit on chairs at the front of the classroom. Provide cards showing all four operation symbols and two sets of large number cards labelled 0–20. Ask one child to choose a number, another to choose an operation (symbol) and a third to choose a number. The class should then complete the calculation using individual whiteboards and, on the count of five, show you their answers. Compile a table showing the calculations and functions. Add a challenge by including a second function: for example, 7 × 8 – 9 = ? Older children could include decimal calculations.
- Play *Guess the Number.* Say, 'I'm thinking of a number. If I add 3 to it, I get 8. What's my number?' Show children the question as a calculation: 8 = ? + 3, or use a number line. When children understand how to find the operation, add difficulty by having two functions: for example, multiply by 5 and subtract 3. You could present this as, 'I'm thinking of a number. I multiply it by 5 and subtract 3. The answer's 22. What's my number?' Ask children to work in pairs to write at least five similar examples. Put them together and make a class book of puzzles for children to do, or display them on the working wall.

Further Activities

- Working in pairs, and using cards from 1–20 and the operation cards, ask children to make their own functions. Emphasise the importance of working backwards to check a calculation. For example, '7 × 6 + 5 = 47. Subtract 5 and divide 42 by 6 = 7'. Pose problems such as, 'I'm thinking of a number. I multiply it by 6 and add 5, making 47. What's my number?' Children should respond with the function. For example, 47 – 5 = 42, or 42 ÷ 7 = 6. Ask children to devise similar problems for each other to solve.
- Provide children with laminated Back to the Start-type function machine problems (see the display). For example, 12 – 8 = 4 + 8 = 12.

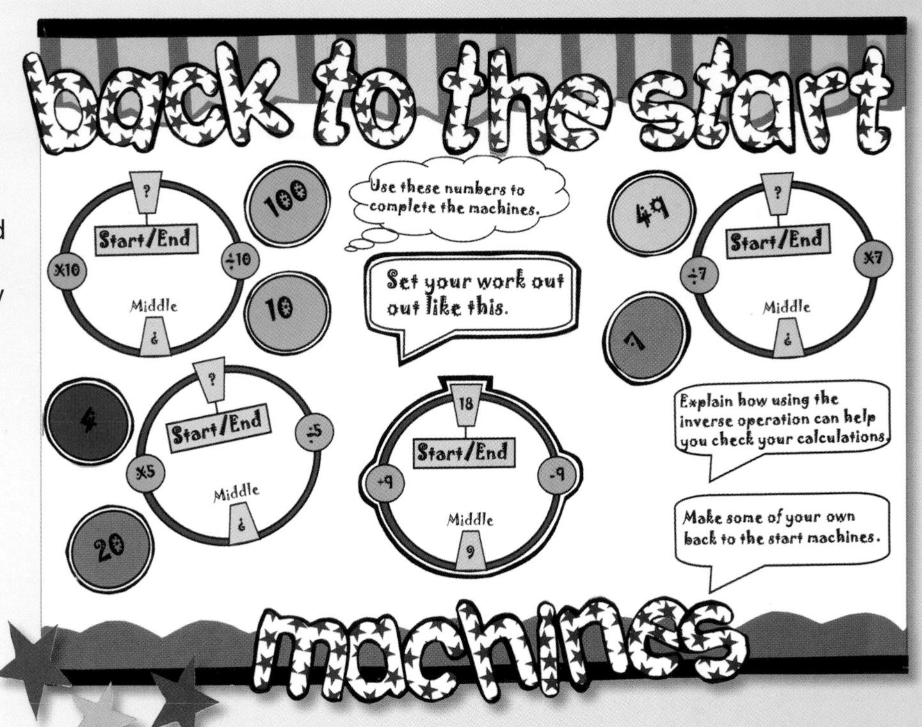

Working Wall

- Ask children to use number cards to fill in the hoops displayed on a board, demonstrating clearly the inverse operation. For example, start with 100 then ÷ 10 = 10, × 10 = 100. (If the numbers are attached with Blu Tack, you can change them on a regular basis, adding challenge by including decimals). Ask children to make their own Back to the Start machines, and display them around the working wall.
- Produce an In and Out machine that clearly demonstrates inverse operations.

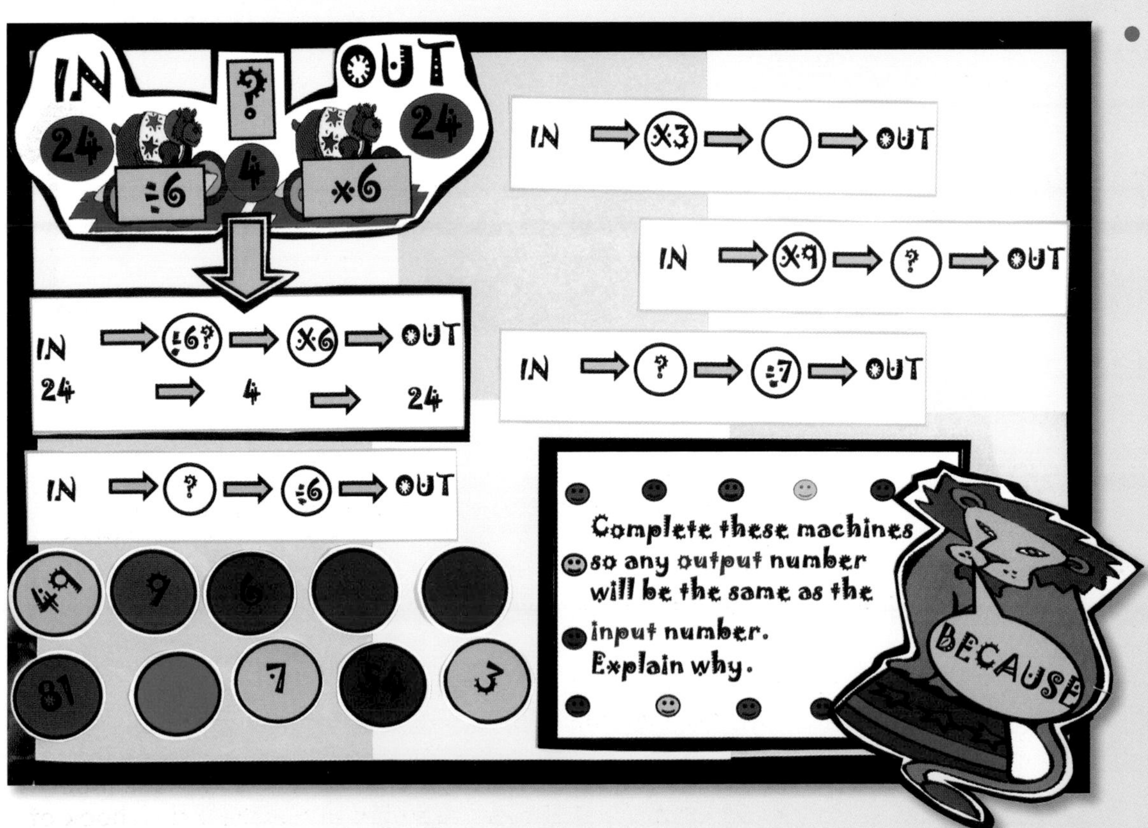

- In a problem solving context, children pick a number from the coloured balls that will ensure that the final output number is the same as the original input number. For example, for the 'In' number, choose the golden ball 3, then decide on the correct function: × 6 (3 × 6 = 18), then ÷ 6 = 3: the original number. Children have to apply their knowledge of multiplication and division tables to answer the questions. Attach numbers and operations with Blu Tack so that the calculation can easily be changed. Add challenge by using negative numbers and decimals.

The Theatre

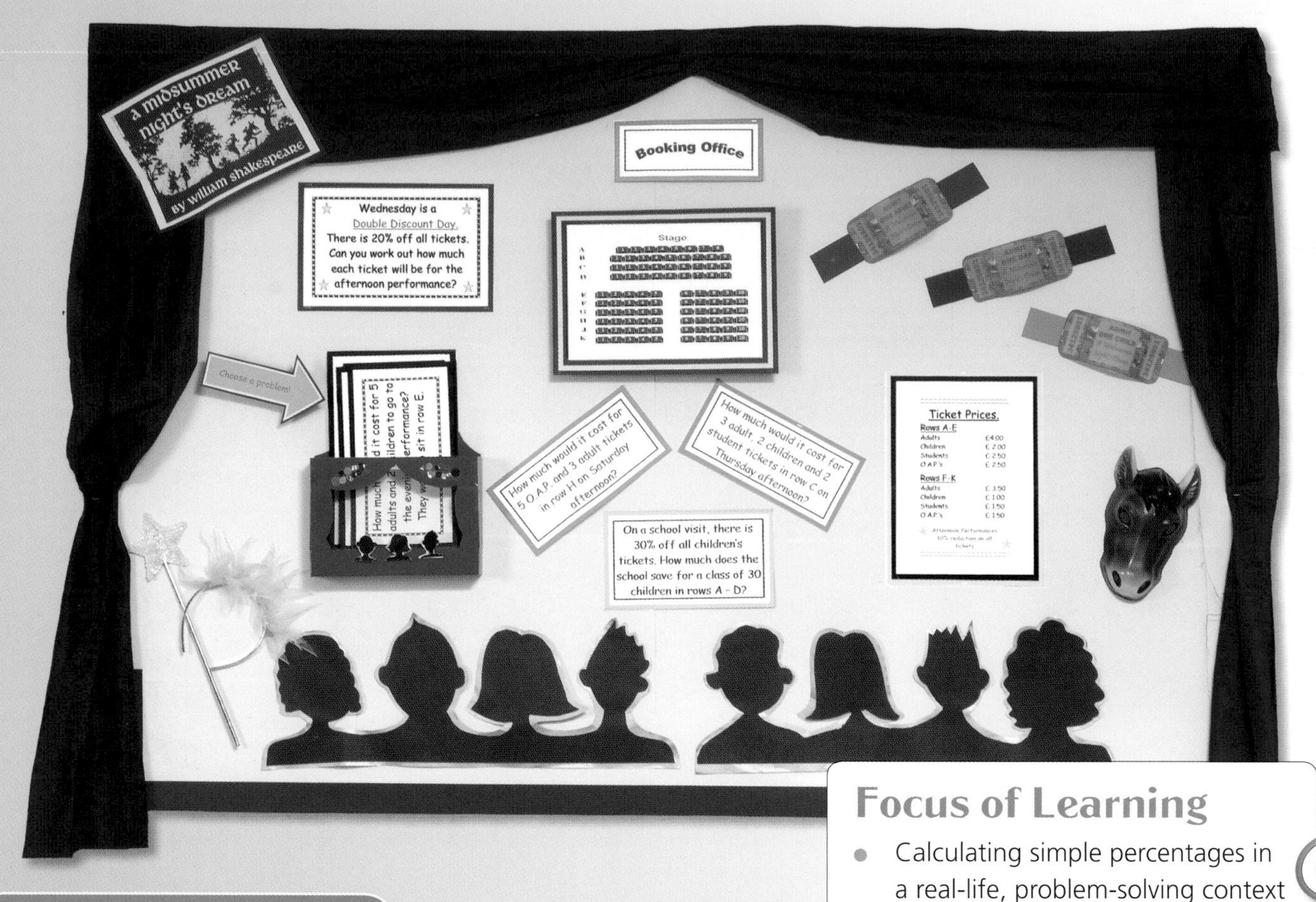

Focus of Learning

- Calculating simple percentages in a real-life, problem-solving context without using a calculator

Art and Display

1. Use curtain material, preferably velvet, to simulate theatre curtains. Cut out silhouettes of children to create an audience. Display a range of items that relate to the play that children are studying, or to school productions.
2. Prepare a plan of a theatre and ask questions relating to percentages of seats, and prepare a table showing the costs of tickets. Add a 'Double Discount Tickets' label. Change the discounts regularly. Prepare a small card theatre to house a range of worded problems.

Starting Points

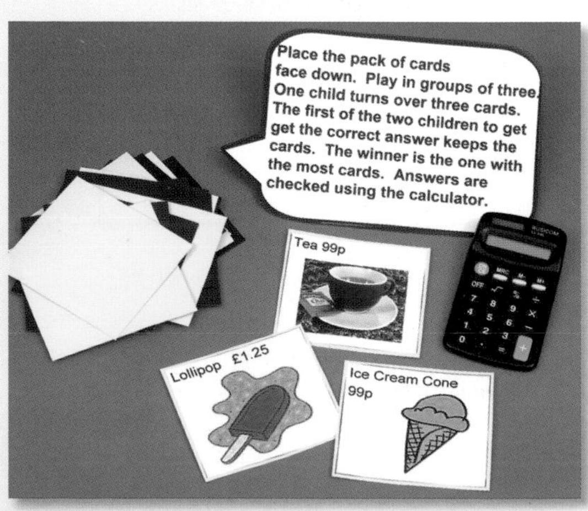

- Play the *Chair Game*. Make a set of percentage and fraction cards covering 0–1: for example, 10%, 25%, 0.5, 0.75, 0.8. Arrange five chairs in a row at the front of the class. Invite five children to pick a fraction or percentage card out of a hat and then sit in the order they think is correct on the chairs. Alternatively, the class can decide where the child should sit, giving reasons. The correct order for this group would be Chair 1: 10%, Chair 2: 25%, Chair 3: 0.5, Chair 4: 0.75, Chair 5: 0.8. The first number out of the hat may be 10%, and the child has to decide if they think it will be the lowest number and then sit on the first chair. Ask probing questions, such as, 'Which percentage/decimal number would come before/after this one?' Children should give reasons for their choice of chair. Add equivalent percentages, decimals and fractions to generate more discussion. For example, ask, 'Where should 75% and 0.75 go? Why?' Alternatively, order the cards on a number line using pegs. This is a game of strategy where children have to use their mathematical reasoning, and the objective of the game is to generate mathematical discussion.

Further Activities

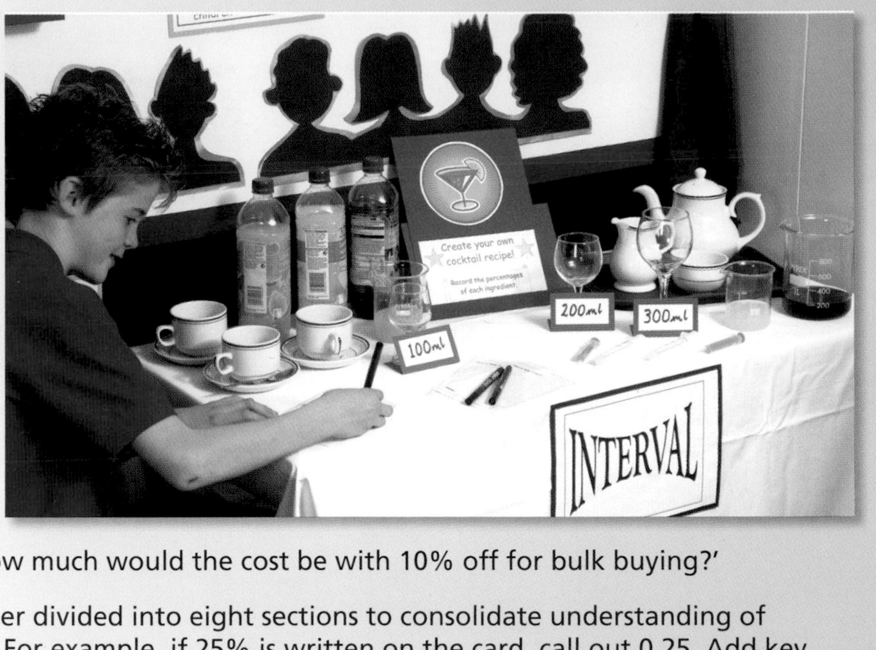

- Create a role-play area, designed as the 'interval area', placed under the theatre display. Ask children to design an interval drink by creating a cocktail recipe. They should use percentages as measurements for their recipe. Children could make their drink using measuring containers and then draw their cocktail recipe with the percentages of juice and water used. More able children could calculate the amounts of juice required for a variety of drinks.
- Price the drinks that children have made at £1.50. Ask children to calculate the cost of drinks for the whole class. Ask, 'How much would the cost be with 10% off for bulk buying?'
- Prepare Bingo Cards on laminated A5 paper divided into eight sections to consolidate understanding of equivalence of decimals and percentages. For example, if 25% is written on the card, call out 0.25. Add key percentages, such as 10%, 15%, 1%, 33.3%, 20%, 25% and 75%, and their equivalent decimals and fractions.
- Create a poster display of appropriate theatre items for children to find the percentage. For example, pose the following problem, 'Programmes for Midsummer Night's Dream cost £5. Tonight there is 10% off, so how much will I pay? What's the difference between the new price and the old price?' Vary the difficulty of the percentages: 10% 15%, 20%, etc. Move on to more difficult percentages: 1%, 12%, 16%. It is useful to ask children to draw a table, with the following headings: 'Old price', '% off', 'New price', 'Difference'.
- Create a set of cards of typical items that can be bought at the theatre: ice cream, lollipops, crisps, ice cream tubs, drinks, and so on, and include equivalent prices. Ask children to calculate the cost of three items and subtract, for example, 10%.

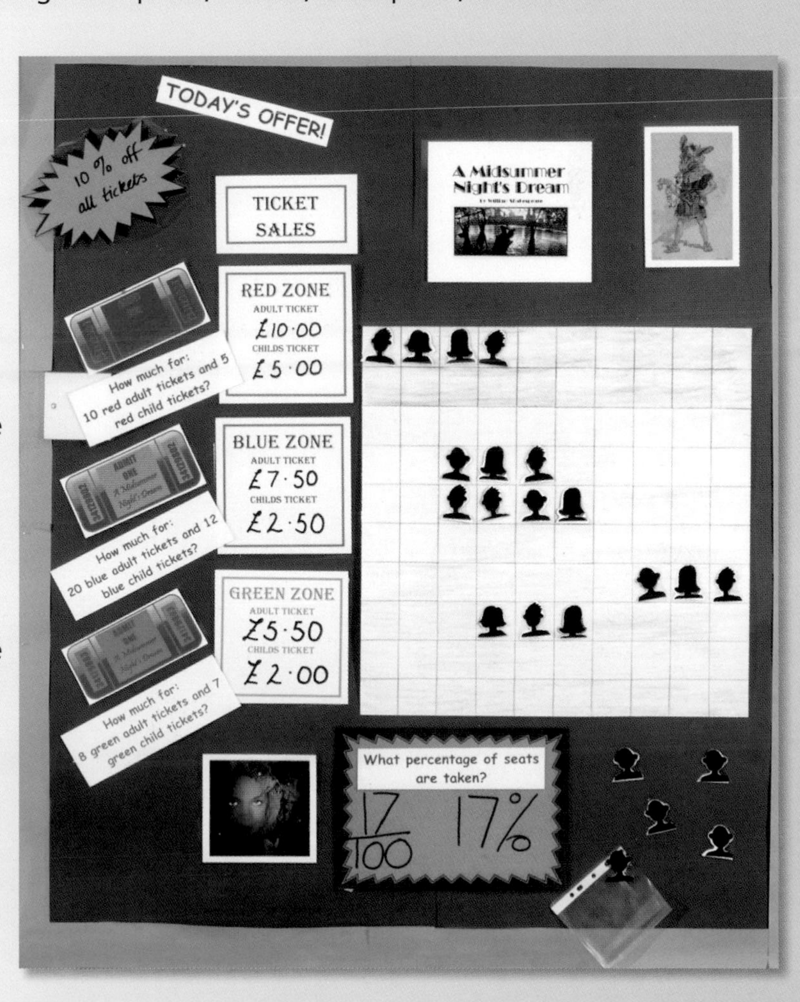

Working Wall

- Make a large 100 square and add different colours for price zones in the theatre for booking seats. Create a set of silhouettes of children and place in a plastic wallet to generate percentage questions. Add labels with ticket prices, and devise a variety of questions relating to tickets prices. Add a label for 'Today's Offer', which is 10%. Use this display as a role-play area for booking theatre seats and calculating the cost of tickets for families and groups of children with percentages deducted. Change the percentages on a daily basis. Use the internet to demonstrate booking theatre tickets.

Cross-curricular Links

- LITERACY, ART and IT – Children can write reviews of the play studied and then display them. They can also write programmes, and design and make them using IT.

Christmas Fractions

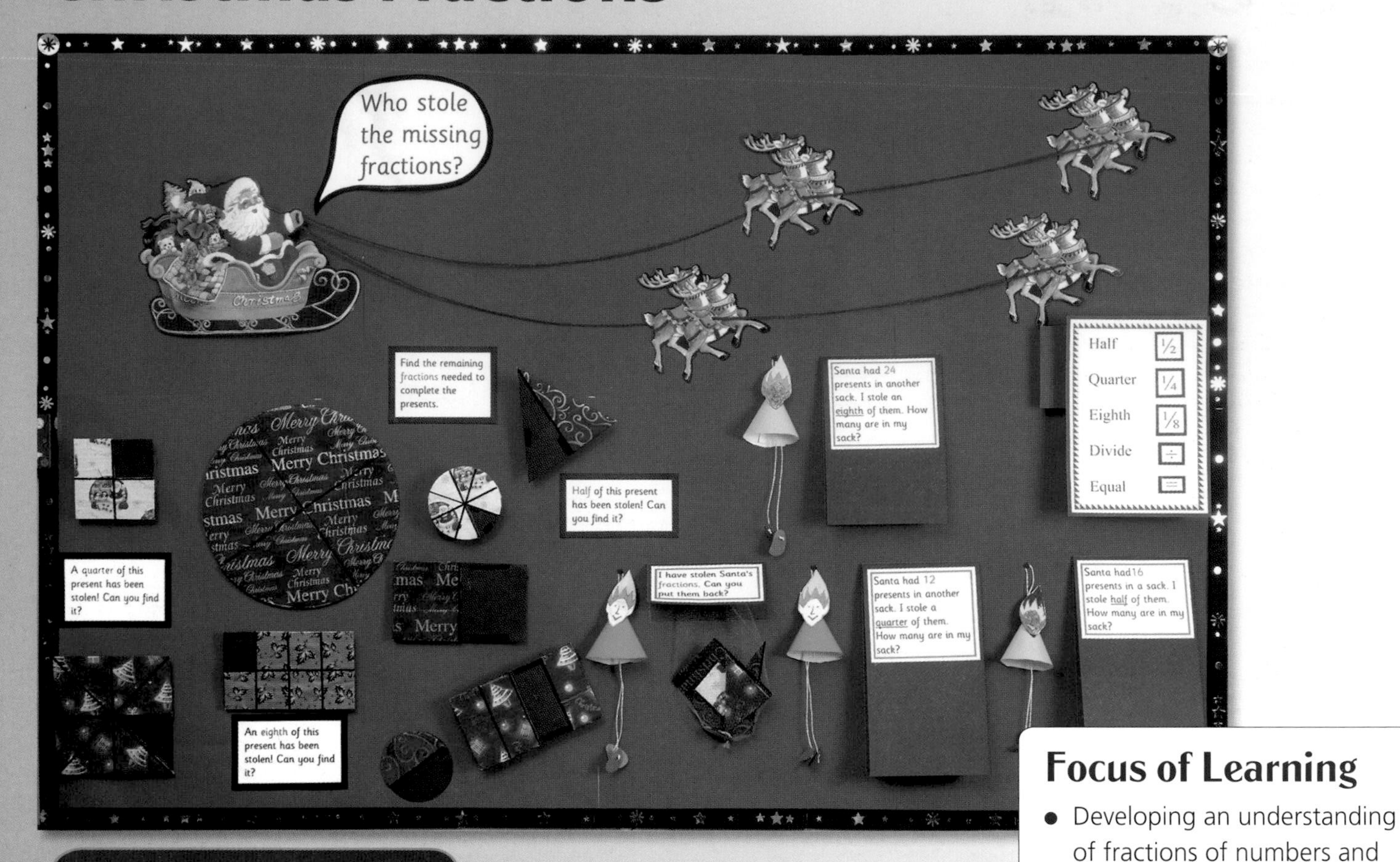

Focus of Learning

- Developing an understanding of fractions of numbers and shapes

Art and Display

1. Produce Santa and his sleigh by using Clip Art and wool for the reins.
2. Create a series of parcels with fractions missing: ¼, ⅛, ½. Add more challenging fractions, according to children's ability.
3. Make elves (with string legs and cone bodies) that have stolen the fractions from Santa's parcels. Ask questions relating to the fractions, such as, 'Santa has 24 presents in a sack. An elf stole an eighth of them. How many are in the sack?' Hide the answers under the red flaps.

Starting Points

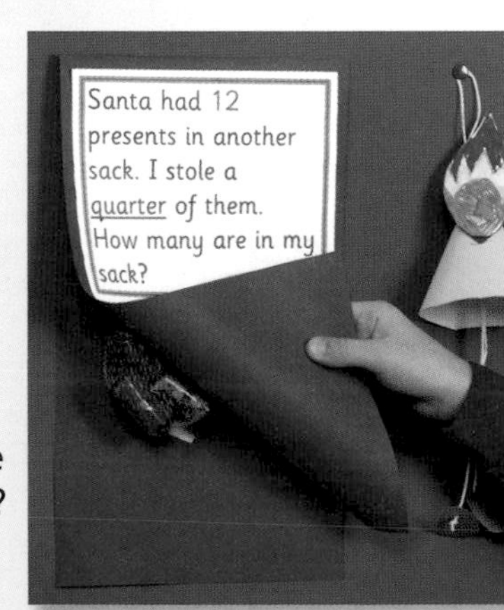

- Ask children to count fractions that bridge a whole one. For example, three-quarters, four-quarters (which is one whole), one-and-a-quarter, and so on. Use a counting stick to demonstrate this.
- Give groups of four children a paper plate with two biscuits on each plate. Ask how the biscuits can be shared. Next, put a sandwich on the plate and ask how the sandwich could be shared. Ask, 'When we share by four, what's the fraction called? How do we write it?'
- Give children a laminated strip of card on which they estimate one-quarter, one-half, three-quarters, and so on. Ask questions such as, 'How much more is needed to make a whole one?'
- Give children templates of parcels with missing fractions, such as eighths, quarters and halves, of varying sizes and shapes. In pairs, children should decide what the missing fraction is and label it. They should justify their reasoning. For example, 'I think ¼ is missing because the parcel is divided into four and there are three pieces left. ¾ + ¼ = one whole one.' Provide a variety of parcels and questions.

Further Activities

- Give pairs of children a small box of raisins or sweets. Ask how they can divide the raisins equally between two. Stress the importance of counting all the raisins and dividing by two by 'grabbing a group', then counting the raisins in the group, not sharing them. If there is an odd number, explain that the one left over is the remainder. Provide 'R' or 'leftover cards' for children to label these remainders.

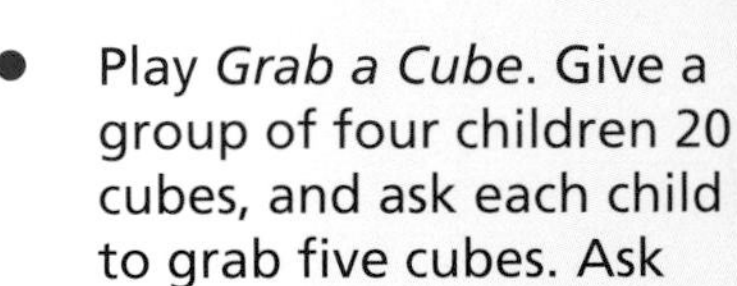

- Play *Grab a Cube*. Give a group of four children 20 cubes, and ask each child to grab five cubes. Ask what fraction of the 20 cubes each of them has. They could record their answers on individual whiteboards. Repeat the activity with different numbers of cubes. When introducing remainders, add extra cubes. Ask children to say what the largest remainder could be and why, and what other numbers could be remainders. Ask, 'I'm dividing by four. What's the largest remainder I can have? Why?' Repeat with different numbers.

- Play *Scoop a Fraction*. Give pairs of children objects to count, such as multi-link cubes, counters or pennies. They should scoop up some of the objects, then count them and decide if they can be divided by two by putting the objects into twos and counting the groups of two. Use the same process to check if 20 can be divided by 4: for example, $20 \div 4 = 5$. Pairs of children could record their work on whiteboards. The first pair to find five fractions is the winner. To make it more challenging, ask them to find thirds, sixths and larger numbers.

- Give children a set of fraction strips, with each strip marked individually into halves or quarters, eighths, thirds and sixths. Children investigate, for example, how many quarters make a half, or eighths make a half. They can then use their strips to compare which two fractions will make, for example, three-quarters. Give children the opportunity to add their fraction to a whole one, saying $1\frac{3}{4}$, $2\frac{1}{4}$, $2\frac{3}{4}$, etc.

- Create a Santa's Workshop interactive display. Give children templates for Christmas crackers, stockings and scarves. They can follow the instructions for colouring in different fractions on the instruction cards.

Working Wall

- Give children four strips of different-coloured A4 paper, cut lengthways. Ask them to fold them in half and label as halves, then fold in half again and label as quarters. A further challenge could be to fold the paper into eighths and label. Ask questions such as, 'How many eights in one-quarter?' These strips can be stuck onto a strip of A4 card and displayed on a display board.

Cross-curricular Links

- **ART** – Use fractions practically when cutting and folding paper sizes, measuring quantities of glitter, pieces of ribbon, string, and so on, in art lessons and as part of this topic.

How Much Is That Doggy?

Focus of Learning

- Developing strategies to support answering worded problems

Art and Display

1. Produce a shop-window display showing a cat and dog, with prices for cans of food. Prices can be changed to increase the challenge of the problem.
2. Add a series of questions relating to the cost of feeding the animals. Use Blu Tack for attaching questions and objects so that the display can be reused to present a different problem.

Starting Points

- Use the display as a starting point for problem solving. Ask questions such as, 'How much would it cost to feed the cat and dog for two days, a week, a fortnight, a month, a year?' Children could use whiteboards to write down their answers. Ask questions that involve children using the items on the board.
- Present children with a calculation: for example, 56 – 7 = ?, and ask them to use the correct vocabulary to answer the question orally: '56 – 7; 56 subtract 7; 56 take away 7. What's the difference between 56 and 7? What's 7 less than 56?', and so on.

Further Activities

- Ask children to make their own worded problem from a simple calculation: 56 – 7. For example, 'John had 56 stickers. Molly had 7 stickers fewer than John. How many stickers did Molly have?' Wherever possible, use visual models to support problem solving. Try with 48 – 21. For example, 'John is 48 and his daughter is 21. What's the difference in their ages?' Use a number line or bead frame to model difference.
- Support children in developing their explanation skills by providing them with a 'because' card to structure their answers. 'I know this is a subtraction question… because…'
- Supporting children in deciding what is the correct operation to use is always a challenge. One helpful way is to give them a set of ten mixed questions, written on separate cards, using the operations addition, subtraction, multiplication and division. In groups of four, ask children to classify the questions according to the operation required. They should explain to their groups that the question is, for example, 'X because…', using the 'because' card.
- Create challenges that involve using problems with one or more steps and one or more operations. Ask children to choose one or two problems to solve.
- Reading and understanding mathematical vocabulary is essential to supporting children in answering worded problems. Children could make a book explaining the mathematical vocabulary to be used.

Working Wall

- A problem solving frame on the wall supports children in developing a clear structure for problem solving. Present children with a worded problem and attach it to the problem solving frame (it is worth using Blu Tack so that you can reuse the frame). Before the lesson, attach paper to the other sections of the frame so that you can write in the information required to answer the question. Leave the information on the display for a few days as a prompt for children. Then change the problem.
- Provide children with individual laminated problem solving frames to support them with problem solving. Give them a printed problem for them to stick in the appropriate place on their individual frame. Frames that are enlarged to A3 can be very useful for group work, particularly when children have difficulty with being systematic in recording and working out what the calculation requires them to do.

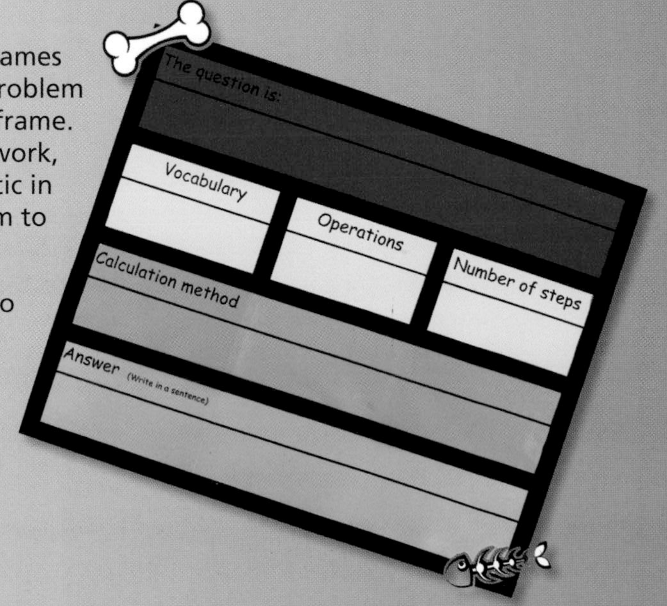

- Ask children to generate a list of strategies that can be used to support solving worded problems. These might include:
  - reading the question twice
  - underlining the important words
  - choosing the appropriate calculation
  - estimating the answer
  - calculating
  - checking
  - evaluating (asking, 'Is this the most effective method?').

That's the Way To Do it!

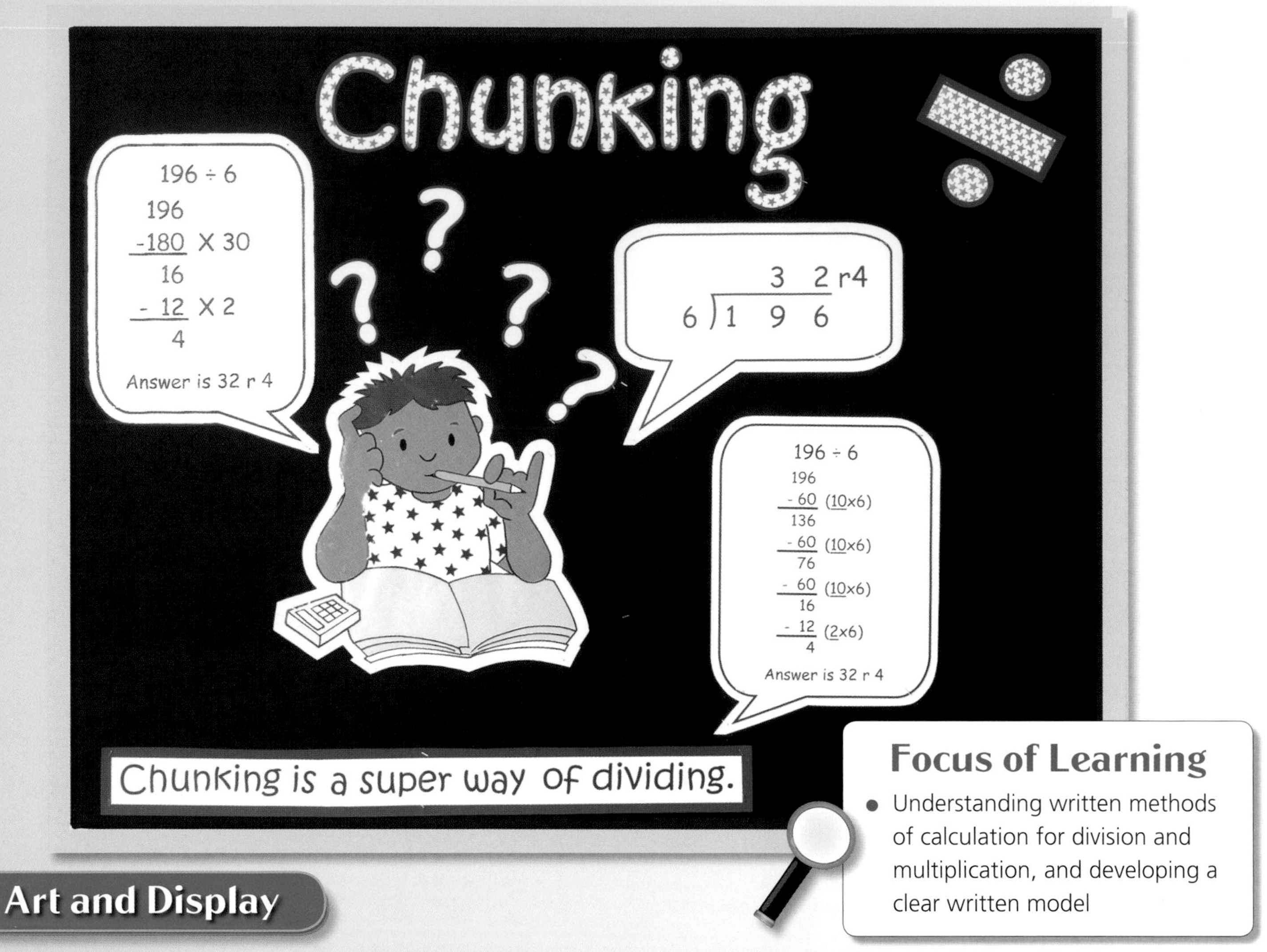

Focus of Learning

- Understanding written methods of calculation for division and multiplication, and developing a clear written model

Art and Display

1. Create a display to show your school's preferred method of chunking for division. In this example, the school's agreed policy showed chunking of 10 × 6 as ten lots of six. Other schools prefer a different method of chunking, such as 6 × 10, meaning six, ten times. It is important that there is agreement and consistency throughout your school about which model to use. Children find this type of prompt very supportive and use it continually throughout lessons. It also supports parents in being clear about the models of calculation that their children are using in school (see page 70 for more ideas for parents).

Starting Points

- Play *Leftover Bits*, by drawing a number line from 0–30 on the board. Write the calculation 25 ÷ 4. Demonstrate that 25 ÷ 4 = 6 remainder 1. Using whiteboards, ask children to show you the largest number, then all the numbers, that could be a remainder for the calculation. Refer to the wall display and ask the same questions. This helps to ensure that children don't leave a remainder that is larger than the number they have divided by, which is a common mistake. Ask questions such as, 'I'm dividing by 8. What's the largest remainder I can have? I'm dividing by 23. What's the largest remainder I can have?' Then explain why.

- Play *Space Invaders*. Give every child a calculator. Key in a number: for example, 5 678. Children then try to guess the number by asking, for example, 'Is there a six in the number?' You answer 'Yes, 600' and subtract 600. The children then add 600 to their display on the calculator. They then ask, 'Is there a five?' You answer, 'Yes, 5 000' and subtract it. The children then add this to their display. When children guess the final number, you will have no numbers on display and they will have 5 678.

Further Activities

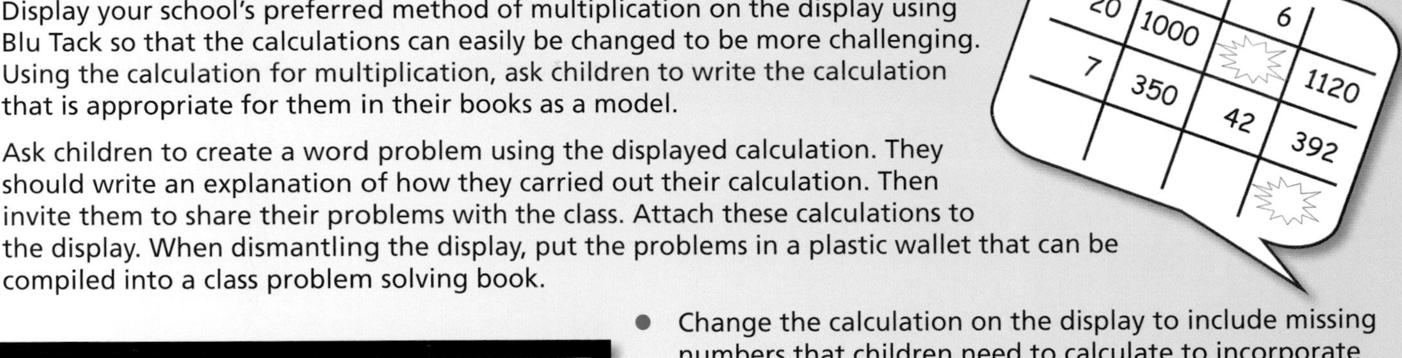

- Display your school's preferred method of multiplication on the display using Blu Tack so that the calculations can easily be changed to be more challenging. Using the calculation for multiplication, ask children to write the calculation that is appropriate for them in their books as a model.
- Ask children to create a word problem using the displayed calculation. They should write an explanation of how they carried out their calculation. Then invite them to share their problems with the class. Attach these calculations to the display. When dismantling the display, put the problems in a plastic wallet that can be compiled into a class problem solving book.

- Change the calculation on the display to include missing numbers that children need to calculate to incorporate a problem solving approach to the written method. Ask children to explain how they found the missing numbers.
- Give children the digit cards 1–10 or three 1–6 dice. They can use these to multiply and divide numbers of digits appropriate to their level of ability.
- Make cards with examples of partially completed or incorrect calculations for children to complete or correct. Being 'teacher for a day' is a very popular activity. Ask the 'teachers' to tick the correct answers and to explain why a calculation is incorrect. They then write a correct model of the calculation and explain what the 'child' needs to do to get the calculation correct next time. This activity generates mathematical discussion.
- Give each child a calculator. Ask them to key in, for example, 3 + + then = . Ask children to continue to press the = sign: the calculator will add on three each time. This will support them in understanding that multiplication is continuous addition. Differentiate the number keyed in by ability.
- Demonstrate place value by using four digit cards: for example, to create 5 796. Put four chairs at the front of the classroom, then ask four children to arrange themselves so that they make the largest/smallest number possible. For the largest number, children repeat the numbers as 9 765. Ensure children always give the full value of each number. This activity supports an understanding of partitioning when using the grid method of multiplication.

Working Wall

- All of the displays featured in this theme can be used as working walls.

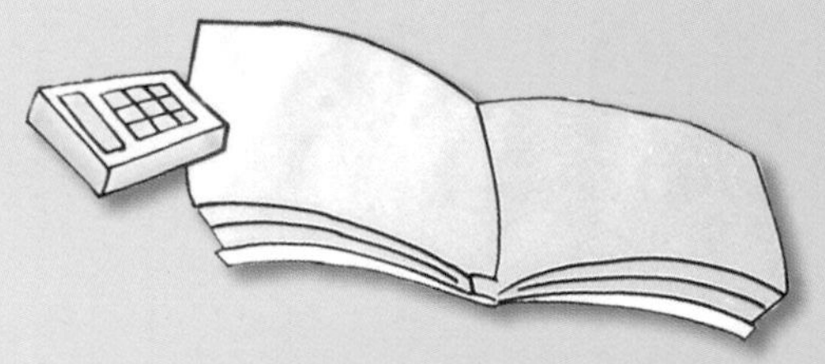

Football Crazy

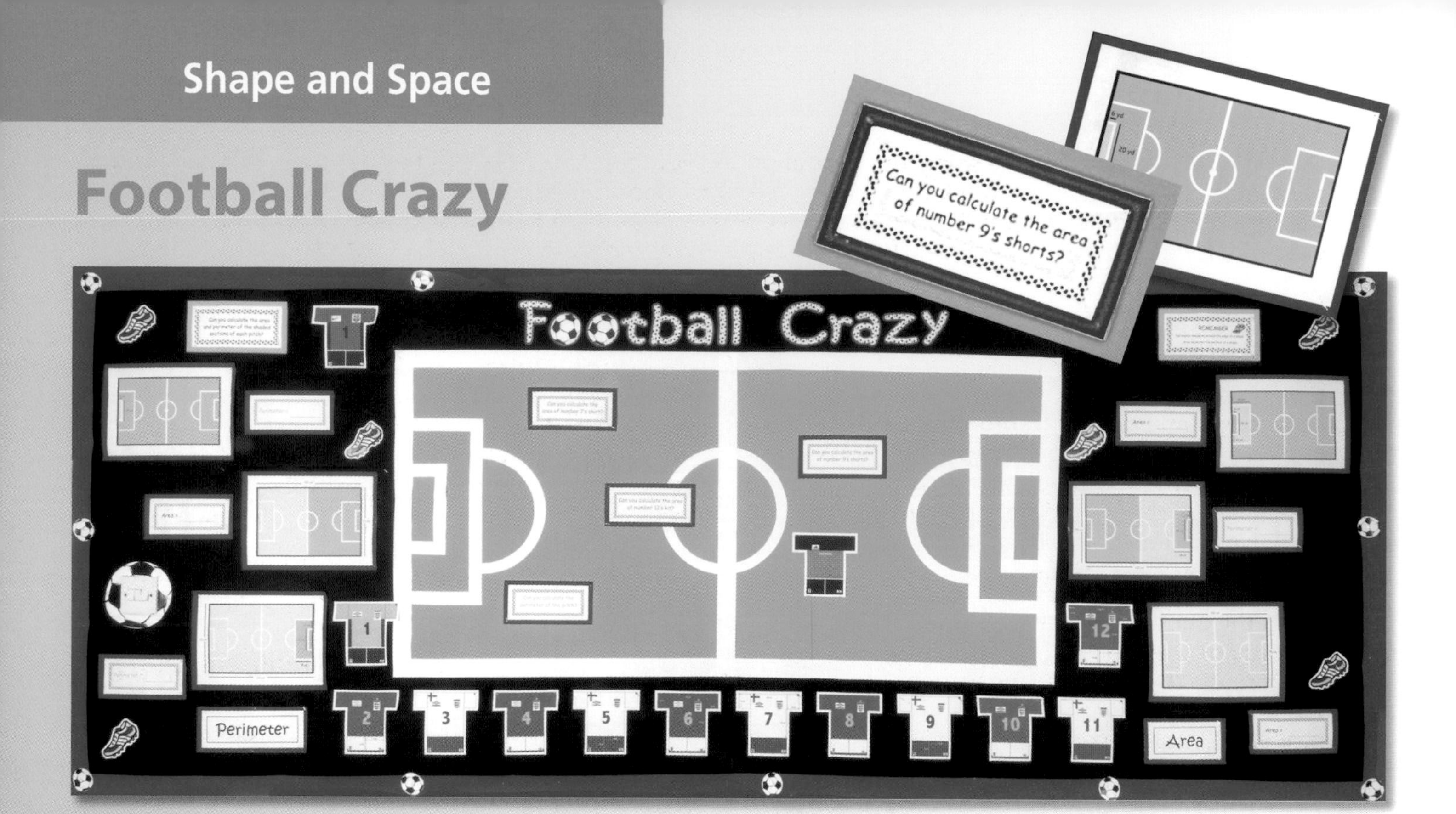

Art and Display

1. Design and make a football pitch, adding the appropriate dimensions.
2. Add the vocabulary 'perimeter' then 'area' as you work through the theme.
3. Explain that the formula for the perimeter of a rectangle is 2 × (L + B), and for area it is L × B (you may prefer to use 'width' instead of 'breadth').
4. Make two sets of card football shirts. Children will later find the area and perimeter of these shapes. Adding stripes and badges can add to the challenge of the calculation. For example, 'Find the area of the shirt not covered by badges.'

Focus of Learning

- Calculating the area and perimeter of regular and irregular shapes
- Knowing that area is a measurement of surface, and perimeter a measurement of length

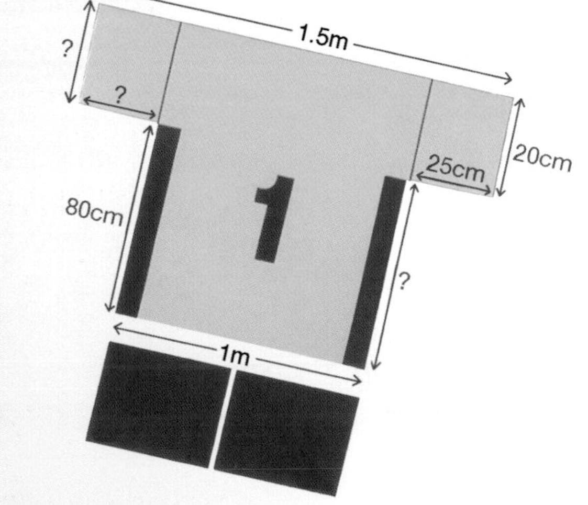

Starting Points

- Take children onto the school field or playground and ask them to walk around the perimeter of the football pitch or netball court. Explain that perimeter is a measurement of length. Point out the word 'rim' hidden in the word perimeter, to help children to remember that perimeter is around the edge (or rim) of a shape.
- Give children geo boards and elastic bands. Explain that the elastic band is the perimeter and children can then measure the perimeter of the shape they create with the boards.
- In the playground, draw large squares or rectangles with chalk. Children can then walk around the perimeter. Give them strips of paper 1m long and hold or tape them onto the perimeter to measure the shape.
- Tape together four metre-long sticks and measure $1m^2$. Extend to $4m^2$, and so on, as a way of clearly illustrating area.
- Give children long lengths of elastic. Working in groups of six, ask four children to make a rectangle or square. The fifth and sixth child should walk around the perimeter of the shape, then measure it.

- Give children examples of the shirts and pitches on the Football Crazy display, then ask them to measure the perimeters. Draw attention to the formula 2 × (L + B) to measure the perimeter of a rectangle. For example, the formula would be 2 × (5m + 4m) = 18 metres. Make calculations more challenging by using decimals. For example, for a rectangle of length 5.4m and breadth 3.2m, the perimeter will be 2 × (5.4 + 3.2) = 17.2m.

Further Activities

- Give groups of children a variety of different-sized boxes and ribbon. Demonstrate how to measure around the top of a box with ribbon, and then measure the ribbon with a tape measure. Ask children to do the same. You can also give children trainers and ask them to use ribbon to measure the soles.
- Using the above boxes, ask children to find the area of the top and one side. Using the formula (L × B), a box measuring 20cm × 8cm would have an area of 160cm^2. More challenging work could involve children finding the area of the sole of the trainer.
- Design and make a photograph frame to display children's favourite football player. Ask them to measure the perimeter of the photograph and design the frame to fit it.
- Give children a carpet tile or a piece of paper 20cm × 20cm. Ask them to run their fingers over the tile/paper, explaining that area is a measurement of surface. Distinguish area from perimeter, which is a measurement of length. In groups of six, ask children to put their paper or tiles together and state the area of the squares.

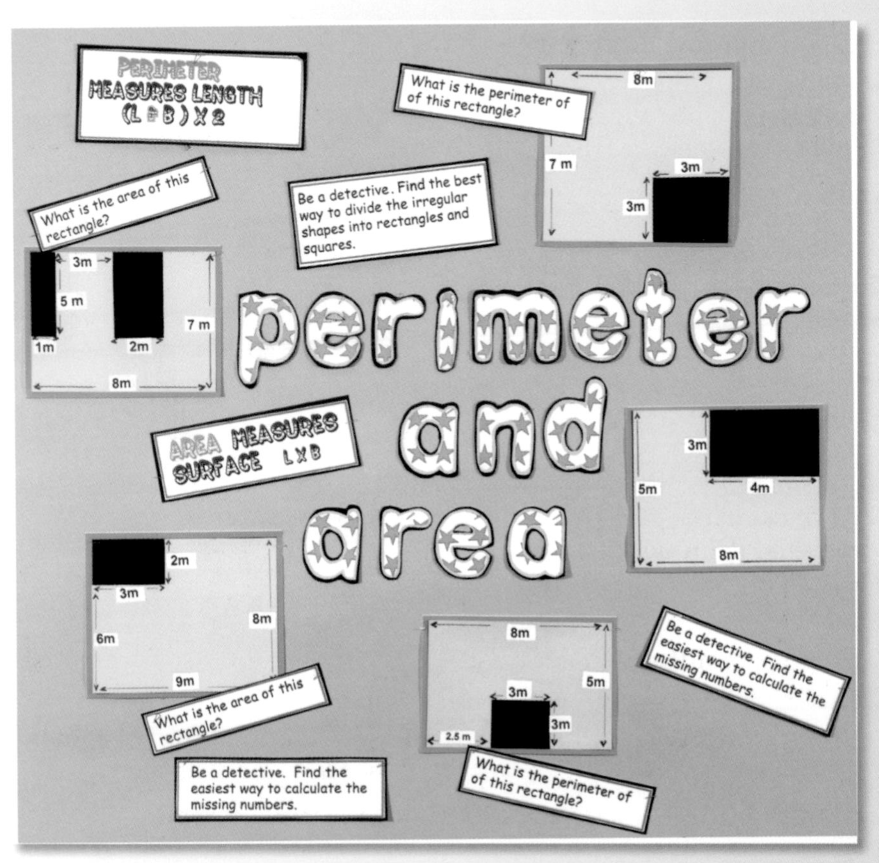

Working Wall

- Assemble a working wall to support children in finding areas and perimeters. Give children a set of small question marks to attach to the measure they don't know. Encourage them to add perimeters by starting with a blob and crossing off the sides they have added. Give them football shirts, with measurements added.

Cross-curricular Links

- DESIGN AND TECHNOLOGY – Each child makes one 10cm × 10cm square for a patchwork quilt, and these are sewn together. This supports the understanding that area is a measurement of surface.
- ART – Design and make a football shirt for a favourite toy or as a motif for a cushion.

Wild Things

Art and Display

1. Create a row and column grid labelled A–F horizontally and 1–6 vertically. Explain that grids are usually made of squares and are spaces in between the grid lines.
2. Provide a key for children to add the appropriate features.
3. Ask children to describe the position of the objects. For example, 'The telephone box is placed on (C, 1).'

Focus of Learning

- Understanding column and row grids

Starting Points

- Using a grid marked A–F on the horizontal axis and 1–6 on the vertical axis (PowerPoint or a projection on a whiteboard makes it easier to see), ask children to place objects on a grid reference. For example, the monster is at (A, 6). Stress to children that the grid reference is the space between the grid, and that they should always read the reference horizontally then vertically. This can be described as 'down the corridor and up the stairs.'
- Play *Changing Places*. Arrange a group of chairs 5 × 4, then ask children to sit on them. Label the first horizontal row A–E, and the first vertical row 1–4. Direct children with questions such as, 'Can whoever's sitting at (B, 3) change places with whoever's sitting at (C, 4)?' Don't use children's names as this gives too much away. Once children understand how the game works, give those who are watching instruction cards for 'changing places'. For example, 'If you're sitting at (B, 1), change places with (C, 4).' Alternatively, using the same grid references, children follow instruction cards to place features from the Wild Things display on chairs. For example, say, 'Place the wild thing on (D, 4).'

Further Activities

- Create a papier-mâché model of an island with a grid labelled A–F horizontally and 1–6 vertically. Place objects on the island for children to record using their whiteboards. Make cards giving the coordinates of objects to be put on the island. Ask children to work in pairs and put objects on the grid, then write sentences to explain where the object is. If you don't have time to make an island, use a m^2 sheet of card cut into the shape of an island and draw mountains, hills, trees, and so on, then laminate and add grid references.

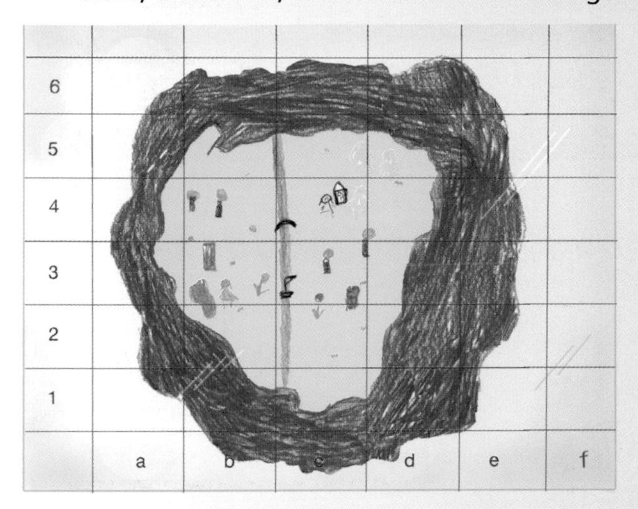

- Ask children to draw the objects and features in their numeracy books, then add an acetate or tracing-paper grid. Children should draw a key showing the exact positions of objects and record the grid references.
- Display examples of children's work in the classroom. A useful activity is to cover the grid references and ask children to find references chosen by the child who has drawn the object. For example, 'What feature can you see at (C, 4)?'
- Play *Find the Wild Thing* in pairs. Prepare 5 × 5 coordinate grids. One child chooses the column and grid position of six hidden 'wild things' and then colours in the position on the grid, keeping the position secret by covering their work with a piece of paper. They should write the reference on the bottom of the grid. The other child has ten guesses to find the coordinate to match the wild thing. They then swap roles. The child who guesses the coordinates in the fewest guesses wins. Stress the importance of checking references, and reading, writing and explaining the references accurately using the correct language. For example, 'The monster is at grid reference (C, 3).'
- In the playground, mark a 10 × 10 grid or use an empty 100 square, if available. Label A–J on the horizontal axis and 1–10 on the vertical axis. Give groups of children ten cards with grid references to place beanbags or objects from *The Wild Things*.
- Give groups of 4–6 children large sheets of drawing paper. Ask them to draw the island where the wild things live. Make small monsters with them from wooden clothes pegs. Make a grid labelled A–F horizontally and 1–6 vertically. (By using electrician's tape, you can easily remove the line without damaging the picture.) Ask children to make a key for the drawing with grid references.

Working Wall

- Children can use the main display as a working wall to find grid references. Change the features and creatures on the display regularly to create interest. Children's own drawings can be laminated and added to the display.

Cross-curricular Links

- **LITERACY** – Use instructional text to describe the position of objects on a coordinate grid.
  – Write creative stories based on the adventures of the wild things on the island, describing features and creatures.
  – Write clear instructions for playing *Find the Wild Thing*.
- **GEOGRAPHY** – Practice supporting simple map reading.

Battleships

Art and Display

Focus of Learning

- Understanding coordinates

1. Make a 9 × 7 grid and label as shown, using electrician's tape for grid lines.
2. Create an operations room and add the labels, as in the display.
3. Use large-headed drawing pins to attach numbered battleships. Then attach string from the operation room to the battleships to show the location of the coordinates.
4. Add a range of appropriate images relating to World War 2, such as newspapers, photographs of battleships, submarines, and so on. Challenge older children by using more advanced coordinates: for example, plotting a battleship or submarine using four coordinates.

Starting Points

- Demonstrate the terms 'horizontal', 'vertical' and 'axis' using an 8 × 8 grid. Explain that the point 0 is where the numbers start and is called the origin. Explain the X and Y axes, and the importance of reading the X axis (horizontal axis) first. Explain that X comes before Y in the alphabet and that this is an easy way to remember the axis.
- Demonstrate writing coordinates correctly: the X axes is always written first, followed by a comma and the Y axis. They are always written in brackets. For example, use the display to show a battleship at (3, 5). Cover the numbers on the two axes and ask children to work out the coordinates of different crosses.

Further Activities

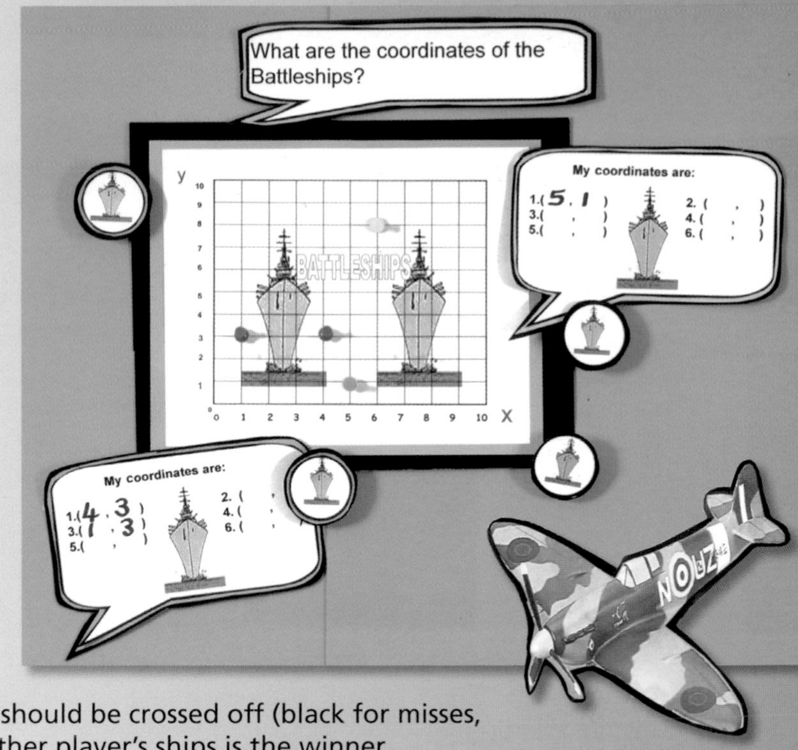

- Explain to children that the X axis can be extended to the left into negative numbers and that the Y axis can be extended down into negative numbers. Explain, 'This type of grid has four areas that we call quadrants. For example, (–4, –4) will be a point in the third quadrant, while (4, 4) will be a point in the first quadrant.'
- Play *Battleships*. Children will now be using a grid to describe the position of something and using the space in between the grid. This will help them to have a clearer sense of the difference. Provide pairs of children with 10 × 10 grids. Each child draws a minesweeper (two squares), a food convoy ship (three squares), a destroyer (four squares) and an aircraft carrier (five squares). Ask children to take turns in guessing the coordinates of their opponent's ships. Any hits (correct coordinates) should be crossed off (black for misses, red for hits). The first person to sink all of the other player's ships is the winner.

Working Wall

- This display (shown) gives tips to support children's understanding of the four quadrants. Provide numbered battleships, submarines and aeroplanes using Clip Art. In pairs, children can place the objects and write down the coordinates. You could also give children individual grids to fill in as a reminder of their work. Ask children to bring small toys to school relating to World War 2 – battleships, and so on – which will help to make the activity more exciting. The working wall can also be used as a role-play area, with children being members of the operation room and recording where the appropriate object is. For example, they write down, 'Submarine sited at (–4, –4)' and add the submarine to the display.

Cross-curricular Links

- **HISTORY** – Include coordinates in a history topic. For example, children could plot the position of pirate ships as part of the Seafarers topic, and the position of the Armada as part of the Tudors.
- **GEOGRAPHY** – Practice map reading.

Polygon Hunt

Focus of Learning

- Identifying the properties of regular and irregular 2D shapes

Art and Display

1. Make a variety of shape characters, displayed as shown. Ensure that you have examples of both regular and irregular shapes, to enable children to understand their properties.
2. Create polygon people, ensuring their feet and eyes match the shape you are studying. For example, the irregular hexagon woman has six sides. You can create images using Auto Shape in Windows XP, or use Clip Art for hands and feet.

Starting Points

- Show children examples of regular and irregular shapes: for example, triangles, hexagons, pentagons and octagons. Ask children to identify the shapes by counting the sides and then holding up the appropriate number of fingers when you say 'Show me'. Emphasise the connection between the number of sides and vertices.
- Use string or elastic to demonstrate the number of sides and vertices on the above shapes. Ask children how many of them are needed to make the shape by holding the elastic at the vertices. For example, a hexagon has six sides and six vertices, so six children will be needed to make the shape. Ask children to hold the string or elastic for the class to decide on the shapes. Ensure they show irregular and regular shapes. They can record their answers on whiteboards. Photographs of this activity make a very good display and support teaching about the properties of shapes.
- Ask children to work in pairs using elastic bands on 5 × 5 pegboards to play *What Shape Am I*? One child makes the shape and the other child explains why it is a particular shape, using the correct language. For example, 'I'm a regular pentagon because all my sides are equal, and I have five sides.' You could assess children's understanding by displaying and numbering a variety of the pegboard shapes, and inviting children to name the shapes.

Further Activities

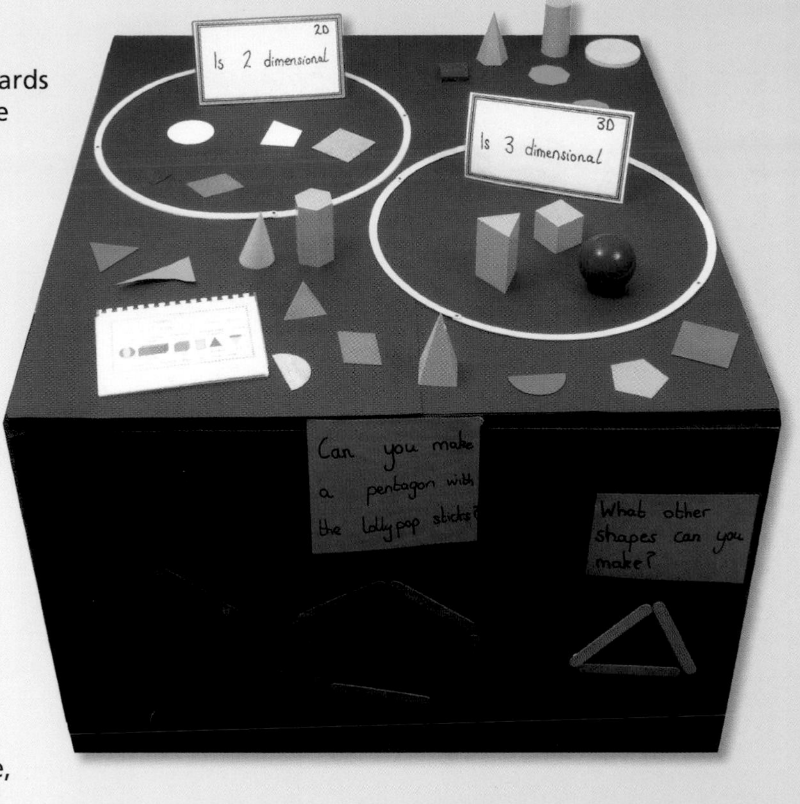

- Play *Shape Bingo*. Give children laminated baseboards with a variety of irregular and regular shapes. Give them definitions of shapes, then ask them to use counters or draw ticks to cross off the shapes. All baseboards could be the same, then all children will shout 'Bingo!' at the same time, or you could differentiate the boards. Make sure you give children shapes in different orientations, and regular and irregular shapes, so that they get used to turning the cards – this will help them to name the shapes.
- Give children a selection of coloured lollipop sticks to investigate the different shapes they can make. Ask children which shapes they can't make and why they can't make them. Then give them different lengths of lollipop sticks, which will highlight regular and irregular shapes. Ask children which shapes they can make now that they couldn't before and to explain why.
- Play the *Yes, No* game. Ask children to sort a selection of regular and irregular shapes in a circle, deciding if they are two or three dimensional. Ask them to place their shape in the appropriate circle and explain their choice. You can also encourage children to draw, colour and cut out their own shapes, which can then be used in the table-top activity.

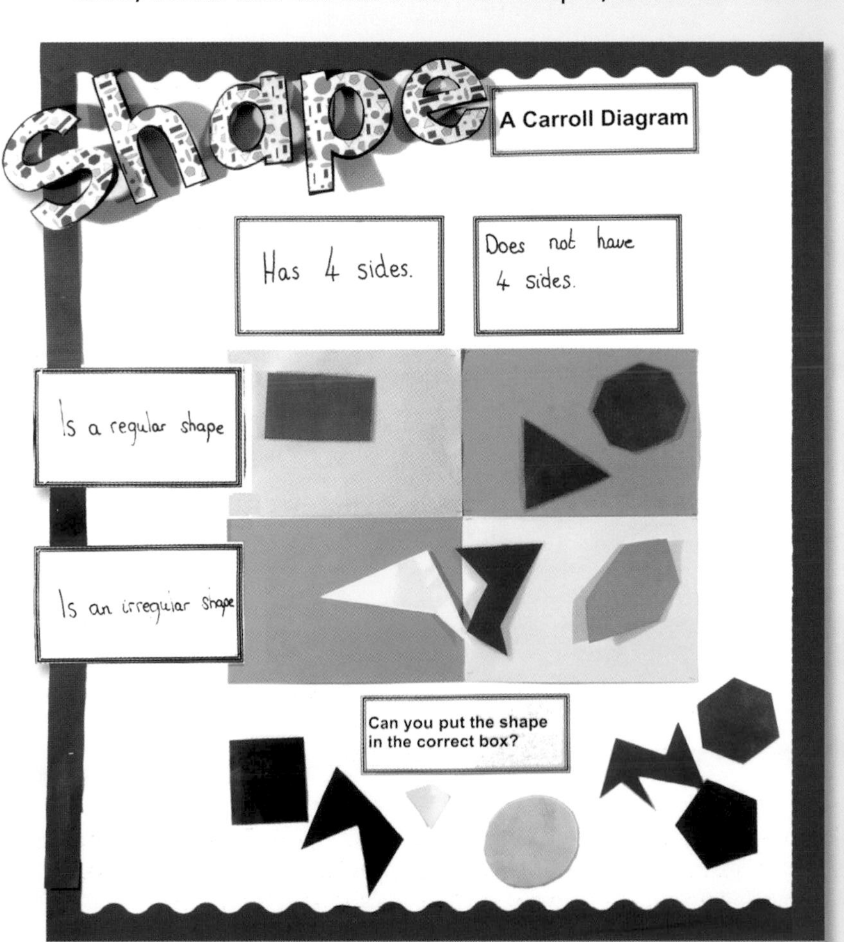

- Give children a copy of the shape characters used in the main display. Ask them to sort them into regular and irregular shapes. Discuss whether the shapes have at least one line of symmetry or none.

Working Wall

- Create a Carroll diagram with a selection of regular and irregular shapes for children to sort.
- Children could use a prepared Carroll diagram and choose their own criteria for other children to sort. Less able children could sort two criteria.

Cross-curricular Links

- ART – Identify the different shapes in the works of artists such as Piet Mondrian (1872–1944), Wassily Kandinsky (1866–1944) and Paul Klee (1879–1940), and geometric shapes in the work of Clarice Cliff (1899–1972). See page 49 for more ideas.

Match the Shape

Art and Display

1. Design and compile a display to encourage children to estimate and visualise nets that will make a cube.
2. Give children 12 different examples of nets for them to choose which will make a cube. They should label their cube with a number and then, using squared paper, draw, cut out and assemble the net.

Focus of Learning

- Identifying 3D shapes
- Designing and making 3D shapes from different nets

Starting Points

- Ask children to bring a box of any shape to class, disassemble it and write their name on the net of the box. Prepare labels with the properties of shapes. Children then place their net by the correct label, according to its properties. Give the nets to other children not named on them, and invite them to name the shape they will make, then assemble them.
- Ask children to identify a variety of 3D shapes. Emphasise that they have three dimensions: length, width and height, and are often classified according to these dimensions. Ask them to use the words 'edges', 'faces' and 'vertices' to describe the shapes. For example, 'My shape has six faces, 12 edges and eight vertices. What is it?' 'I'm a triangular prism because I have five faces, nine edges and six vertices.' Ask a child to deconstruct the shape and check the estimate by counting the properties on the net. Give pairs of children a variety of boxes for them to estimate the number of faces, edges and vertices, then deconstruct the shape and fill in a table showing its name and properties.

- Ask children to visualise a toy box the shape of a rectangular prism. To demonstrate, paint the front, top and bottom of a box yellow, the back red and the other faces blue. Ask probing questions such as, 'How many edges are there where a yellow face meets a red face? How many faces are there where a red face meets a red face?' Ask children to estimate answers, draw them on their whiteboard, then check their answers using a prepared net of a cube.

Further Activities

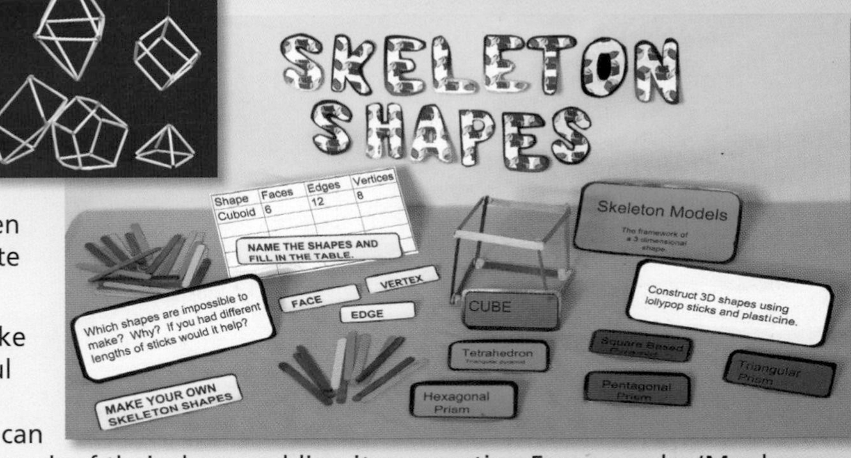

- Give groups of up to six children a variety of small toys. Invite them to make a box for the toy. Give them nets (for reference) to choose from. Children then design their own net and make and decorate the box.
- Give children straws or lollipop sticks to make 'skeleton images' of 3D shapes. This is useful as they can see the number of vertices and edges on their shapes very clearly. Children can then design cards with a drawing or photograph of their shape, adding its properties. For example, 'My shape is a triangular prism. It has five faces, nine edges and six vertices.' Provide a table for children to write in the properties of all the shapes they have made.

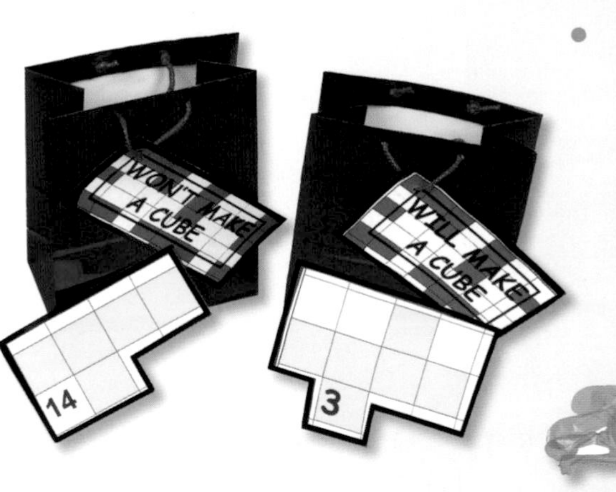

- Give a group of up to six children 14 different nets that may or may not make cubes. They sort the nets into 'will make a cube' or 'won't make a cube', then make the cubes to check if they are correct. Use their findings as a display board. Use the same activity to identify nets of rectangular prisms, triangular prisms, and so on.

Working Wall

- Prepare a working wall that involves children choosing the correct net, the name and its completed 3D shape. Blu Tack the net of the shapes so that children can check if they have chosen the correct shape by folding the net to make the shape. Add a table for children to fill in the relevant facts.

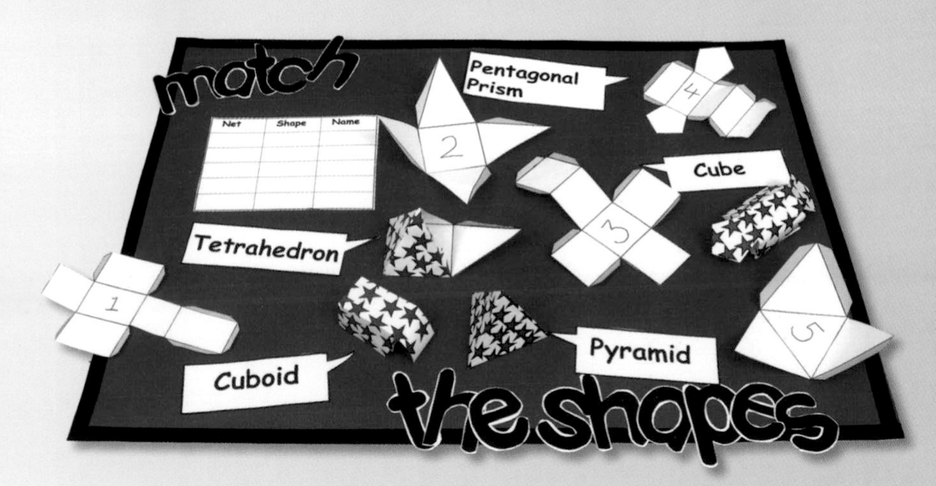

Cross-curricular Links

- DESIGN AND TECHNOLOGY – Use 3D shapes to make a variety of robots. Ask questions such as, 'How many different shapes can you see? How many faces of the small cube aren't glued to the body? How many faces of the rectangular prism can you see? How many edges does a rectangular prism have?' Design a box for the robot and ask children what information they need to start with.
- LITERACY – Design and make a poster to describe the features of the children's robot.

Winter Tile Factory

Focus of Learning

- Tessellating shapes

Art and Display

1. Prepare a Winter Tessellation display. Add tessellating hexagon snowflakes made by folding and cutting paper.
2. Using prepared templates of reindeer, penguins and fir trees, children can colour them, and add wobbly eyes and bobble noses to the reindeer, and wobbly eyes to the penguins.
3. They can attach their tessellating shape to the display, manipulating it so that it tessellates. They can then add labels made from hexagons.

Starting Points

- Explain that, to tessellate, patterns must fit without leaving a gap or overlap at the junction. Show pictures of tessellation around the school, such as bricks, hexagonal tables, roof tiles, floor tiles, and so on. Take children for a walk around the school and ask them to draw all of the shapes that they see that will tessellate.

- Give children a variety of regular shapes such as triangles, pentagons, equilateral triangles and rectangles, and so on. Using whiteboards, ask children to divide the board into two and write 'will tessellate', 'won't tessellate' and sort their shapes. Children could then make a tessellating pattern from their chosen shape.

Further Activities

- Give children a template of a hexagon. Ask them to use a range of different colours cut from magazines or Christmas wrapping paper to make a tessellating pattern/border pattern or decorative photo frame.
- Give children a large selection of scraps of materials, sequins, beads and pom poms, which they can use to make a tessellating blanket for Santa or a scarf for the snowman. They can choose a shape from hexagons, pentagons, squares, equilateral triangles or octagons. They then sew these shapes together or use a piece of card for a base. Cut shapes can then be tessellated and either sewn together or glued onto the piece of card. To give a neater finish to the blanket, use card as a base.
- Children can make Christmas cards, labels and wrapping paper from the tessellating shapes used in the main display: reindeer, penguins, Christmas trees and snowmen. They can use a template to draw around or cut out the chosen shape on card. They can decorate this with sponge and mirror stars, ribbon and coloured paper. Paste the tessellating Christmas shapes onto the Christmas cards and labels.

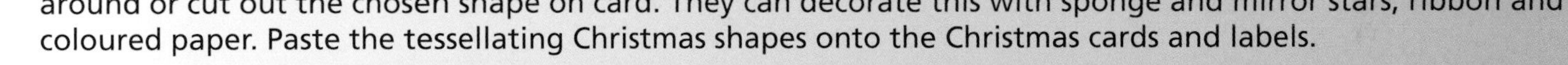

- Children can design and make Christmas wrapping paper using the Christmas templates and a chosen shape: for example, a hexagon. They draw around the hexagon and add the Christmas template within the shape, then decorate. Alternatively, they can make Christmas wrapping paper using IT. First, they can choose a shape from the shape menu: for example, a hexagon, then use the cut and paste function to cover the A4 paper with hexagons. They use the minimise function in Word to minimise the spread, then choose Christmas Clip Art images. The images can be cut and pasted onto the wrapping paper within the hexagons.

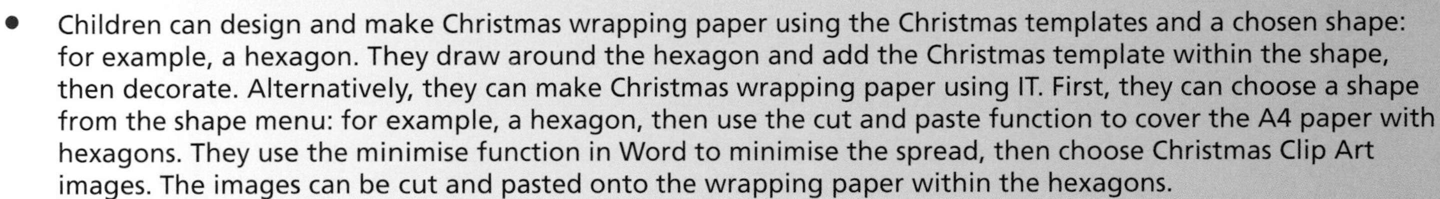

- Ask children to make a tessellating tile. Provide examples of how to make different tiles and cards for them to experiment with. Give them squares or rectangles of card that are $8cm^2$, or 8cm by 4cm rectangles. Cut a piece off one side of the square or rectangle, then stick the other side of the shape. Draw around the shape and make it into a template. Repeat this to create a border pattern.
- Support children to make complex patterns. Using an 8cm square of card, make two cuts in the cards and move them onto the sides of the square. This could make a small tessellating pattern, which could be framed to hang on a wall. Children can use the internet to research artists such as M C Escher (1898–1972) and display some of his artwork. They could write instructions for a friend to follow to make their tessellating tile. Display the best pattern and instructions.

Working Wall

- Using the template for snowmen or fir trees, children can cut out and decorate a shape to create the working wall. They then add it in the correct position to tessellate, using a protractor to check that the joins are 360 degrees without any gaps.

Cross-curricular Links

- **ART AND LITERACY** – Create a class book of tessellating designs. You could also display instructions on how to make them.
- **RE** – Research religious tessellating patterns. Emphasise that the theme of tessellating patterns could be Easter, or holidays, or could be woven into many different topics.

Sensational Symmetry

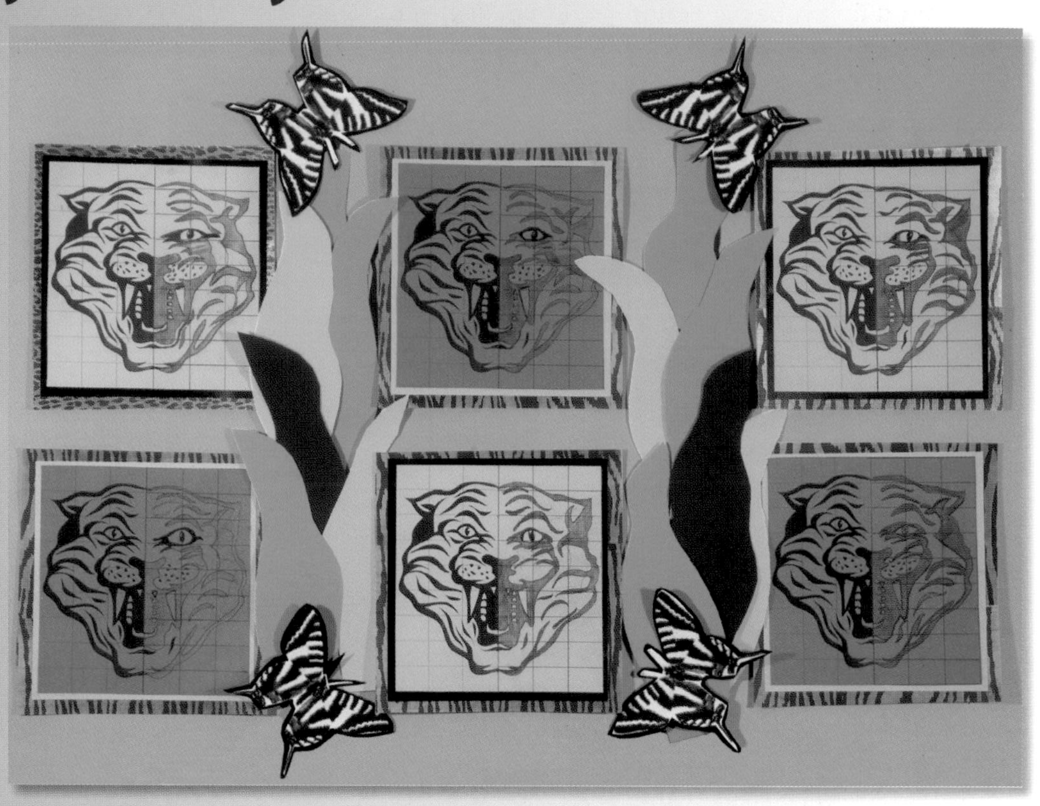

Art and Display

Focus of Learning

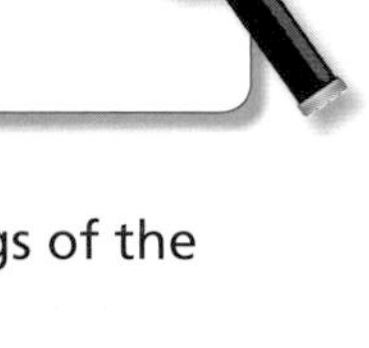

- Understanding symmetry

1. Choose a theme as a starting point for a symmetry display: for example, the jungle.
2. Prepare a background for children's reflective symmetry drawings. The symmetry drawings should reflect the topic you are covering: for example, the Tudors' drawings of the faces of Henry VIII, his wives, Hampton Court Palace.
3. Use a grid to support children when they are creating the drawings, as this encourages detailed and accurate work. The grid also supports children who find it difficult to see the reflection.
4. Give children a drawing of a tiger and an empty grid for them to complete. Emphasise the importance of counting the squares from the centre of the drawing to ensure perfect symmetry. Arrange butterflies around the display and add leaves.

Starting Points

- PE is an excellent starting point for symmetry. Children can sit in pairs, facing each other, with crossed legs. Child A is the leader making the movements, and Child B mirrors these movements. They then change roles. More advanced work can involve standing and mirroring movements, or completing a sequence of movements, ending with a symmetrical still. Photographing the movements and labelling them makes an instant display. With parental consent, cut photographs in half, add a grid, then children can complete the symmetrical picture of themselves. This can extend into area and perimeter as children make photograph frames.

- Give children a 6 × 9 grid, where the symmetry line has been drawn on the fourth vertical line, and give them multi-link cubes or counters in two colours only. Each child can work with one colour. In pairs, one child can place a cube, then the other places their cube so it makes a symmetrical pattern. Children can continue until the cubes have run out. The more intricate the pattern, the better understanding children have of symmetry.

Further Activities

- Give children a variety of halved faces of celebrities cut out from magazines. Give them a mirror and ask them to complete the celebrities' faces. If children are struggling, give them grids to help them.
- Alternatively, choose people from the topic you are studying. In this example, images were used relating to building and to World War 2, but you could choose any historical character.
- Give children six cut-out squares. The aim is to find as many different shapes using all of the six squares, which should be attached along the length of at least one side. Give children mirrors to check for one or two lines of symmetry. Then children can transfer their drawing onto cm^2 paper. Alternatively, use interlocking cubes for this activity but stress that it is the 2D outline of the same shape, not the 3D model, that is the focus.
- Give children a variety of shapes, squares, rectangles, hexagons, and isosceles and equilateral triangles. Ask them to fold the shapes in half and cut a symmetrical pattern. They can then design a pattern on squared paper and transfer the pattern to fabric to illustrate reflective symmetry using fabric paint. Alternatively, the pattern can be completed in cross stitch on fabric.
- Play the game *Symmetry Match*. Give pairs of children a sheet of squared paper divided into two, and give them a pegboard that has been divided into two using an elastic band, pegs and a plane mirror. Child A puts a peg on their side of the board and records it on their squared paper, while Child B puts a peg in the correct position on their side of the board. They record each other's moves. The game continues until the board is complete. Check children's moves for accuracy. A challenge for older children could be to use the four quadrants to place the pegs.

Working Wall

- Children can make witches' masks and glaze them to create a working wall. Faces could have links to World War 2 or characters from history. Encourage children to discuss the following questions: 'Which faces are symmetrical? Which faces aren't?' Ask them to explain why. For example, the witch might have a mole on one side of her face. Extend the activity by explaining that most faces aren't symmetrical. By putting two left sides of a face together, the face will alter slightly.

Cross-curricular Links

- **ART** – Design pottery faces with symmetry.
- **RE** – Research pattern in religion: for example, Islamic art and Rangoli patterns.

Art into Maths Goes

Focus of Learning

- Investigating angles and shapes through the designs of Clarice Cliff

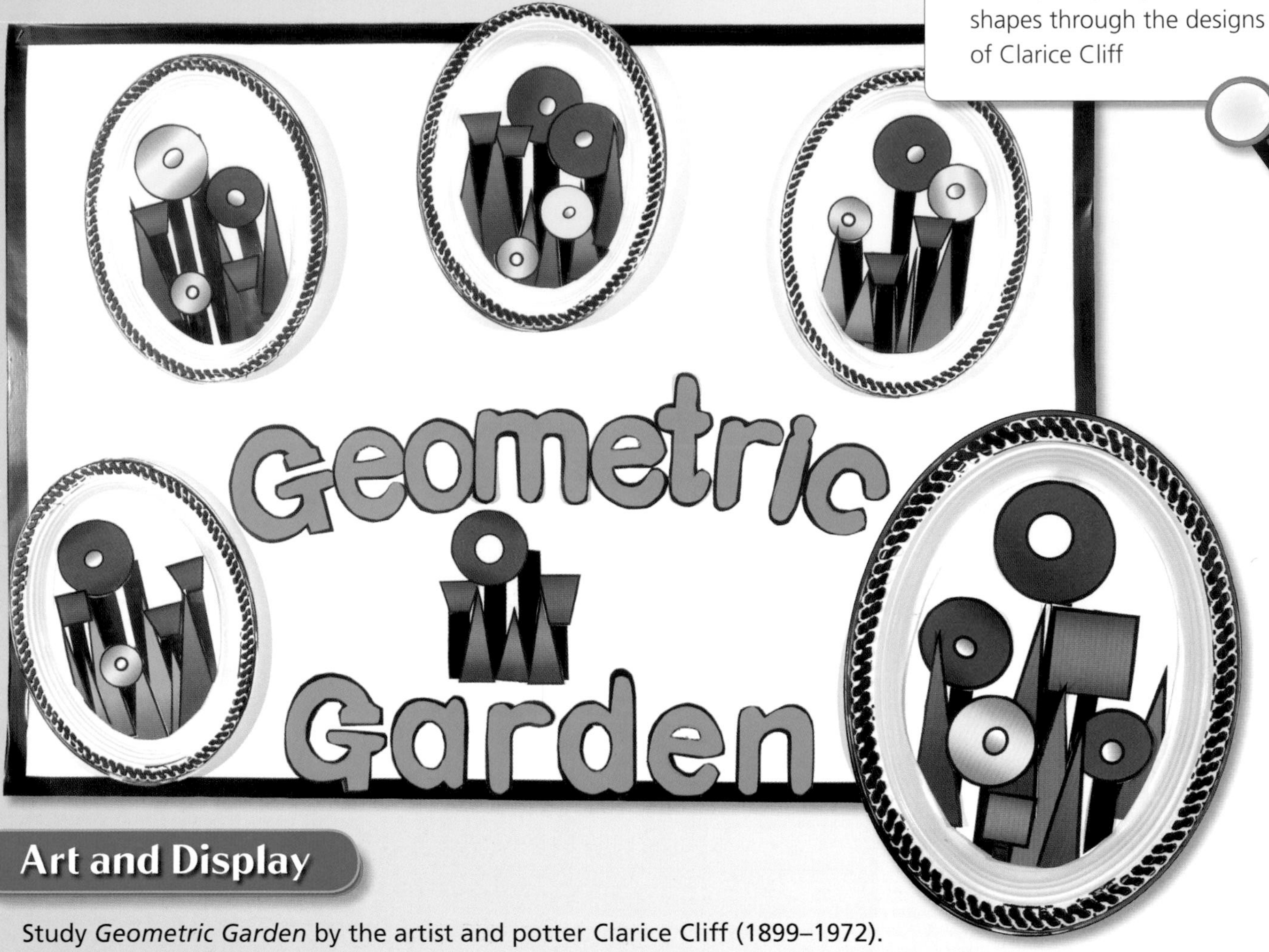

Art and Display

1. Study *Geometric Garden* by the artist and potter Clarice Cliff (1899–1972). Identify the shapes she used.
2. Give children oval paper, which they can attach their design to. Using an art program such as AutoShapes on Microsoft XP, children can select the fill tool, then use the shading option or the line tool to outline the shapes with shading or thick black lines. The shading tool makes the shapes luminous. They can then cut out the shapes and stick them onto the plates. Other options could be to use templates and different-coloured card. Children could cut around the shape and attach it to the plate. They could also add tissue paper to the designs to lend them a 3D quality.
3. As the shapes are attached to the plates, draw attention to the different shapes that are made by overlapping them. Encourage children to count the sides and name the shapes.

Starting Points

- Recap on the properties of the shapes that you have decided to work with. Play *Behind the Wall* by putting a shape behind a cardboard wall and uncovering a small section of it. Ask children to answer questions such as, 'What shape do you think this is and why? What shape couldn't it be and why?' As you reveal more of the shape, ask, 'Do you still think the same or have you changed your mind. If so, why?'
- Play the *Shape Game*. Give children the properties of, for example, a square: 'I have four straight sides. I have four right angles. My sides are all the same length. What am I?' Ask, 'Which property helped you to make the decision? Why?'
- Play the *Finger Game*. Tell children you are an octagon, and ask how many sides you have. When you say 'Show me', children can show you eight digits. A variation of the game is to ask children to draw the shape on their whiteboards, then show you their shapes.

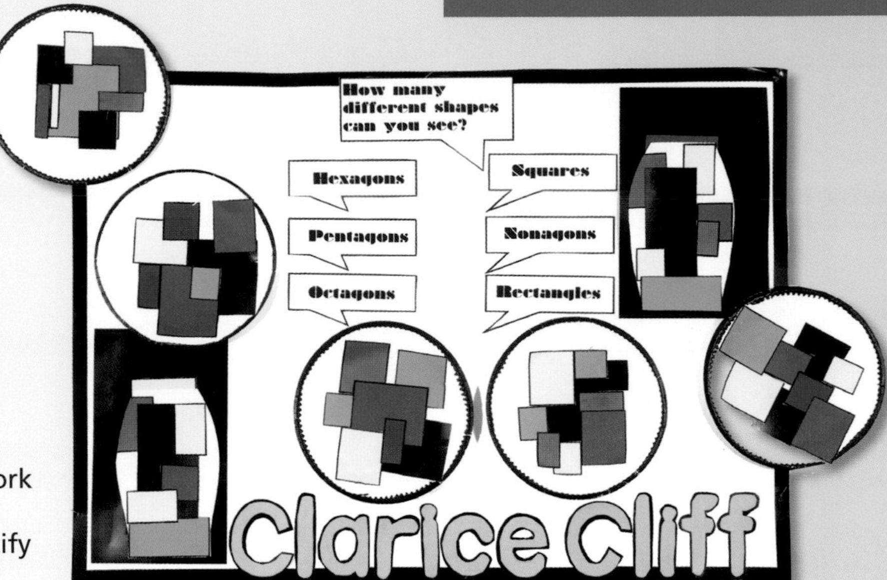

- Ask children to close their eyes and visualise a shape. Say, 'I have four straight sides. Two pairs of sides are equal and all four of my angles are right angles. What am I?' An extension can be to ask children to draw their visualisation before you show them the named shape.

Further Activities

- Give children examples of relevant artwork by Clarice Cliff, such as vases, plates, teapots, and so on. Ask children to identify the shapes they are going to use in their design and write them on whiteboards, together with their properties. They can then write labels identifying the shape and the properties of the shape used. More advanced work could be to identify areas and perimeters of compound shapes, as in Clarice Cliff's *Football* design.

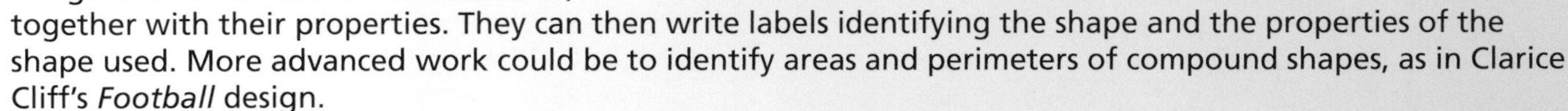

- Using clay, children can design and make their own vase, plate, and so on, in the style of Clarice Cliff.
- Give each child a beaker with a card sleeve on it. Invite children to paint the sleeve on the beaker in the style of Clarice Cliff, using pictures of *Bizarre* and *Geometric Garden* as references. An easier alternative is for children to use shapes they have identified in the pictures, cut them out, stick them onto the beaker and add detail using paint. Glaze using glue and water.

Working Wall

- Give children a selection of shapes: for example, hexagons, pentagons, octagons, squares, nonagons, rectangles. These can be computer generated or children can make them by drawing around templates. As a class, look at the bold patterns that Clarice Cliff used and point out the variety of shapes. Using the shapes in three colours only, children can make their own patterns and stick them onto paper plates and/or templates of vases using different sizes of rectangles and squares. They should overlap the patterns to make a variety of different polygons. Pick a shape and ask children, 'How many sides does this shape have? What is it called? What else do we know about the shape?'

Cross-curricular Links

- LITERACY – Use IT to research the life of Clarice Cliff and the variety of pottery she made and designed.
- HISTORY – Research life in the time of Clarice Cliff, as well as her own life.
- ART – Children find the bright colours and geometric patterns of *Bizarre* and *Geometric Garden* interesting, and can use them to make their own patterns and designs.

Location, Location

Focus of Learning

- Using compass points to mark positions

Art and Display

1. Create a spider's web with three circles, labelled 1–3.
2. Add the compass points N, NE, E, SE, S, SW, W and NW.
3. Position a spider and a variety of different flies on the web, using Blu Tack. Number each fly so children can recognise them easily to state their positions.
4. Add an example as a model for children to position the flies.
5. Create wallpaper using flies for lettering.

Starting Points

- Put the compass points N, E, S and W on the wall in the hall or classroom. Facing north, ask children to turn 90 degrees clockwise (a quarter turn), then ask where they are facing. Vary the questions after making turns of 90, 180, 270 and 360 degrees. A challenge is to ask where children would be facing if they turned 45 degrees after facing north (this would be NE), then 90 degrees from NE (this would be SE).
- Draw a large circle on the playground or hall floor, label the four or eight compass points and mark a dot in the centre. Ask a child to stand in the centre of the circle and to turn, clockwise or anticlockwise, 45, 90, 270 and 360 degrees.
- Check children's understanding of ROARS, which stands for Right angles, Obtuse angles, Acute angles, Reflex angles and Straight lines. Call out, for example 'Right angle', then children should make a right angle with their arms or hands. Make the activity more challenging by stating only the degree of the angle. For example, call out '80 degrees' and see if children can show an angle of almost 90 degrees.
- Give children two intersecting circles, 10cm in diameter, in different colours. Make one cut to the centre of the circle. Intersect the two plates to use as a 'Show me' activity for angles.
- Use a roamer, dressed as a spider, for children to practice their understanding of angles of turn. Add cardboard boxes as obstacles for the roamer to negotiate. Use 360, 45, 50 and 75 degrees of turn. Encourage children to write instructions to guide their roamer spider.

Further Activities

- Give children a location grid, which contains the alphabet placed at different points on it. Ask them to write a message using the points on the grid. They should first write the distance from the centre, then the direction. For example, 'S is at (5, E).' Encourage children to swap their coded message with a friend.
- Prepare an island called Arachnophobia, and add forests, trees, lakes and mountains. Children can also make spiders that live on the island. Ask them to plot the position of the town, features and spiders, and their positions in relation to one another.

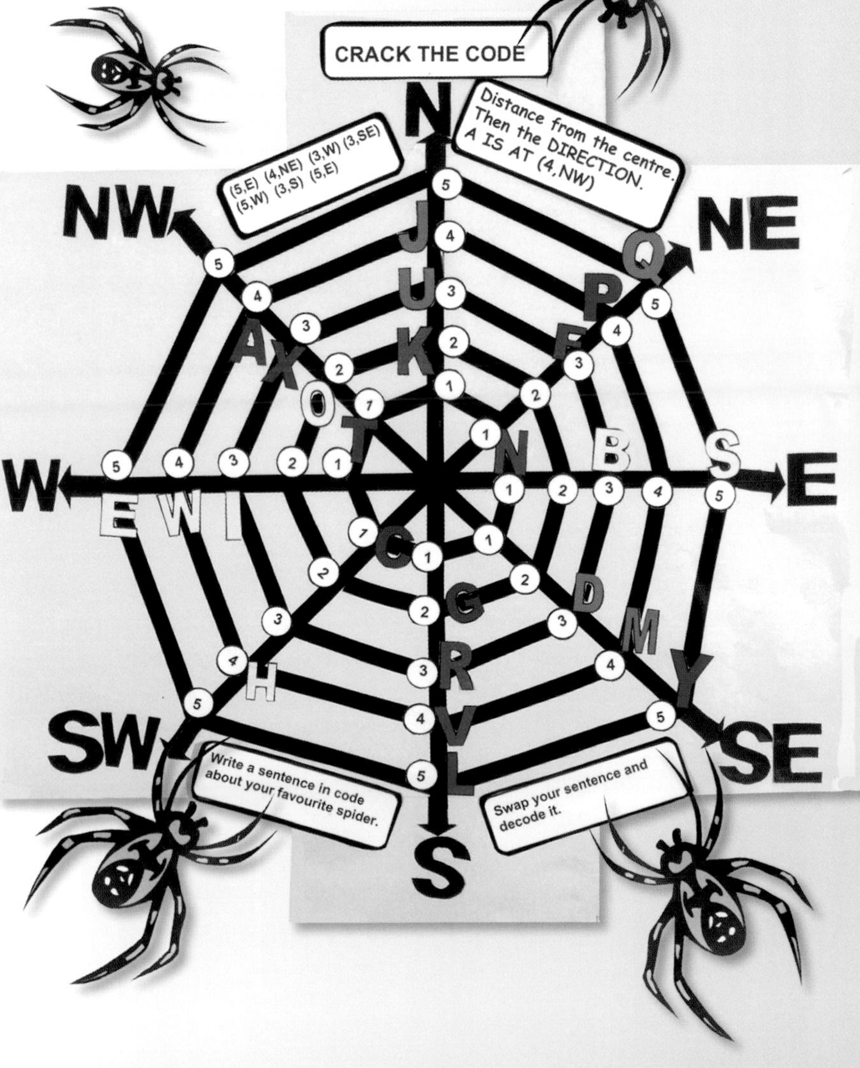

Working Wall

- Create a spider's web, as in the display. Using the web, children can crack the code to find the secret message. They can investigate different types of spiders, then leave a coded message about their spider in an envelope for other children to decode.

Cross-curricular Links

- GEOGRAPHY – Use compass points to enable children to follow directions when reading local maps. They can also draw their own maps and add features using specific points of the compass. Use compass points and turning through angles when studying wind direction using a weather vane.
- LITERACY – Write clear instructions for children to follow when planning a journey. Provide instructions for children to follow to add features on a map.

  – Use descriptive writing to describe the terrain of a journey.

Patterns and Puzzles

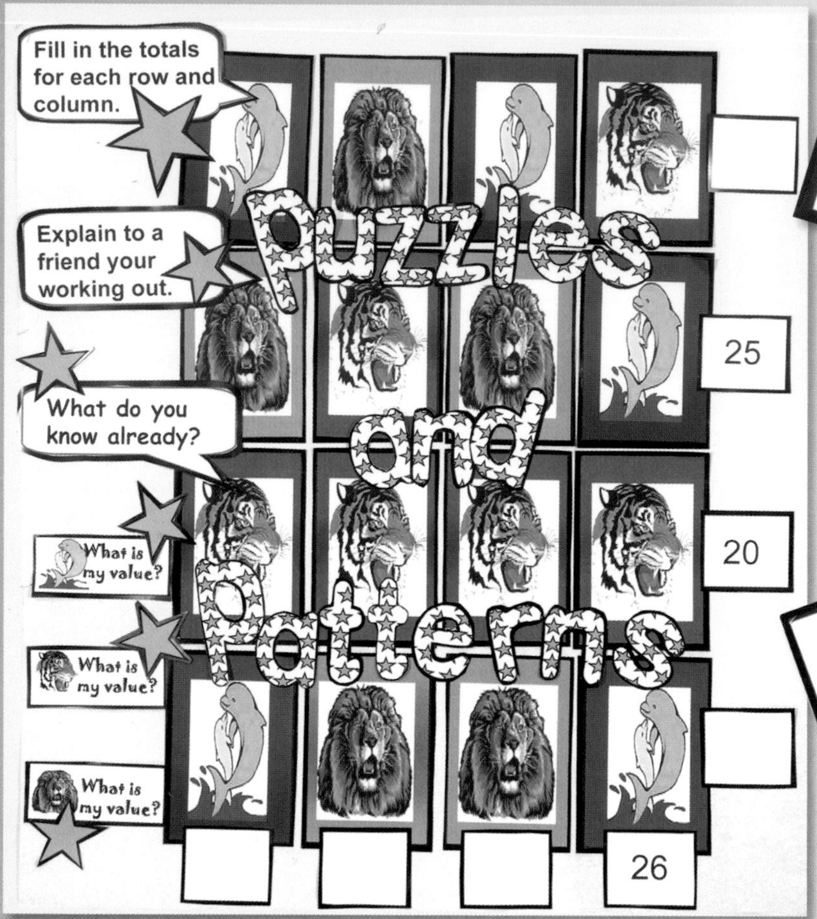

Focus of Learning

- Problem-solving with patterns and puzzles

Art and Display

1. Create a display showing a table with four types of animals, with each animal representing a number.
2. Ask children to use the information provided on the puzzle to answer the questions. For example, 'The tiger stands for 20 because 20 ÷ 4 = 5. The dolphin stands for 8 because 26 – 10 = 16 ÷ 2 = 8. Calculate what the remaining animals stand for. The zebra stands for [add line] because [add line]. The lion stands for [add line] because [add line]. What information do I know? What information do I need to know to answer the question? Which problem solving strategies have I used to answer the question?'

Starting Points

- Ask children to find the missing number in a sequence * 136, 147 * * *. Ask them, 'What information do I need to know to answer the question? How can I work out the difference between the two numbers?' They need to know at least two of the numbers to work out the difference between them, which they can then count on from the next number. For example, the difference between 136 and 147 is 11, so they will need to add on in multiples of 11.

- Give children sequences of numbers to find the pattern. Refer to the display. Give children a 4 × 4 grid showing the puzzle on the display, and demonstrate what a row and a column mean. Ask children to explain what they already know and how they know this. Ask the following questions: 'What number does the lion stand for? How was this number calculated? How does knowing the lion stands for five help with the calculation? What information do you need to help you to calculate the number (for example, doubles or numbers you already know)?'

Further Activities

- Ask children to devise a puzzle for a friend to complete, using a 4 × 4 grid and a symbol for a number. Total numbers must be below 30. When the puzzles have been tested, children can make them into a class book of mathematical puzzles.
- Invite a child to think of a number between 5 and 15. They should keep the number secret. Then ask them to add 10 and give you the answer: for example, 16. Therefore you can work out that their number was 6: 16 – 10 = 6. Now write the child's number on a piece of paper without divulging it or put it in an envelope and continue the calculation. Then ask the child to – 6 (10), × 2 (20), – 4 (16), ÷ 2 (8), – 2 = 6. Tell the child to keep the number secret in their head. Pull out your piece of paper and say, 'Your number was 6!' Try it again with a couple of other numbers. Ask children to work in pairs and check the puzzle. Then try bigger numbers. Ask, 'Does it always work?' Children enjoy 'tricking' their friends.
- Make three parcels: one red, one blue and one green. Explain that the red and green parcels together weigh 7kg, the blue and green parcels together weigh 8kg, and the red and green parcels together weigh 11kg. Ask, 'What's the weight of each parcel?' Children can work together in groups of four to solve the problem. Find out what strategy they used. See if they can explain to the next group of children how they solved the problem.

Working Wall

- Create a working wall, as shown. Ask children to use lolly sticks to continue visual patterns, using small faces to show the joins. Give them a table to complete to show the pattern. More able children could draw the patterns, missing out the first stage.

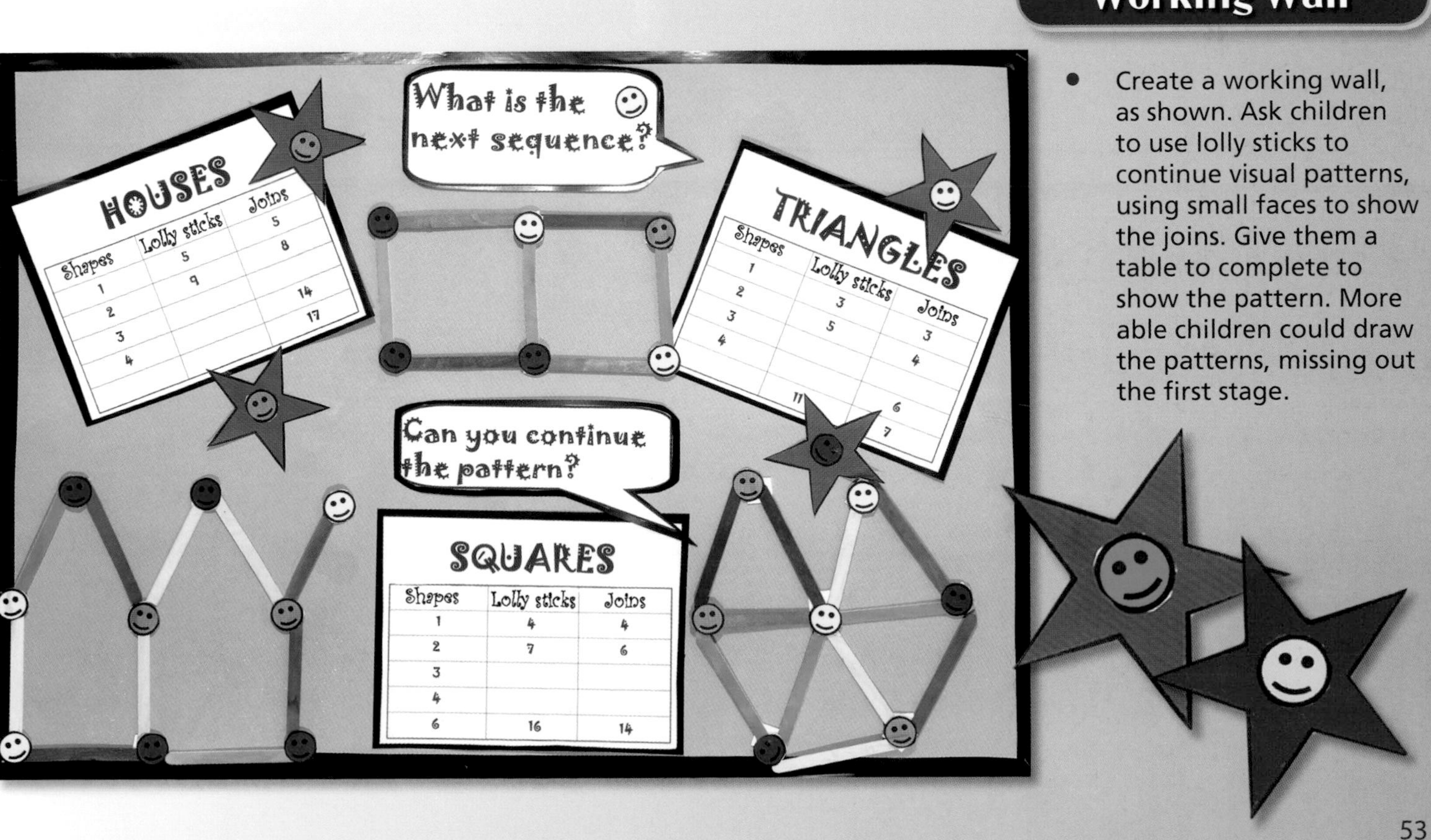

Healthy Eating

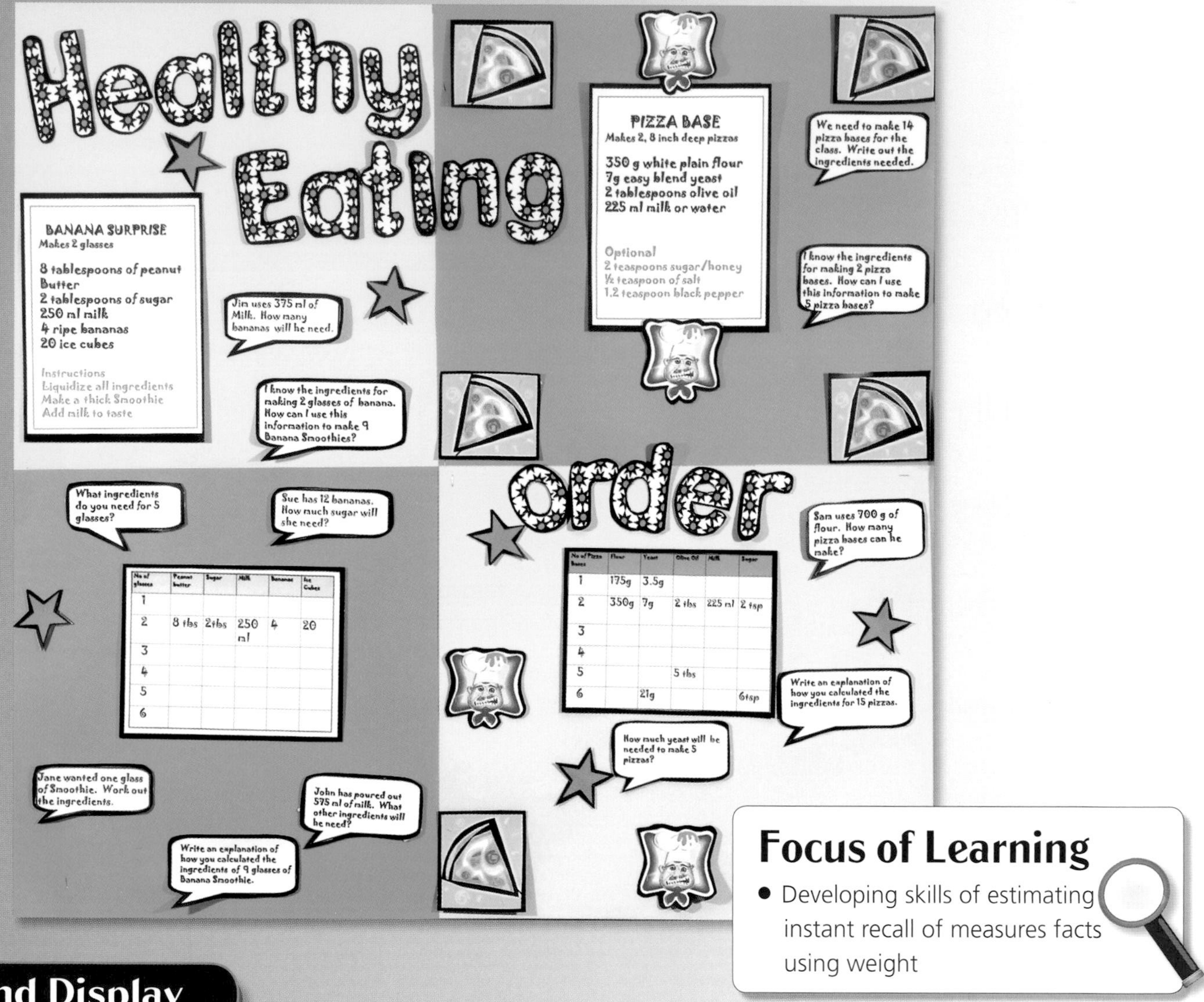

Focus of Learning

- Developing skills of estimating instant recall of measures facts using weight

Art and Display

1. Prepare a board in two colours, using stars for the lettering. Devise two healthy eating recipes: Pizza Base and Banana Surprise.
2. Ask children to choose Clip Art to decorate the display. Add the heading 'Order', where children can order different numbers of pizzas and the appropriate ingredients.
3. Create two tables to support children in calculating the amount of ingredients they would need for different numbers of people. For example, if 350g of flour is needed for two pizza bases, the recipe would be doubled for four pizza bases, and halved for one base. For five people, the 350g would be doubled to give 700g for four people, and the 350g for two people would be halved (175g) and added to 700g, giving the answer 875g. Using a table, and doubling and halving, makes a difficult concept easy.

Starting Points

- Using a counting stick, ask, 'If this end of the stick is 0 and this end is 1,000, what units could we be measuring in? How do you know?' Pointing to halfway on the stick, ask, 'How many grams is this? How do you know? How many grams would be in one-quarter, three-quarters, one-tenth, and so on, of a kilogram?'
- Give children different weights under a kilogram and ask them to order the weights from the lightest to the heaviest. They can record the weight, check using a scale and find the difference between them.

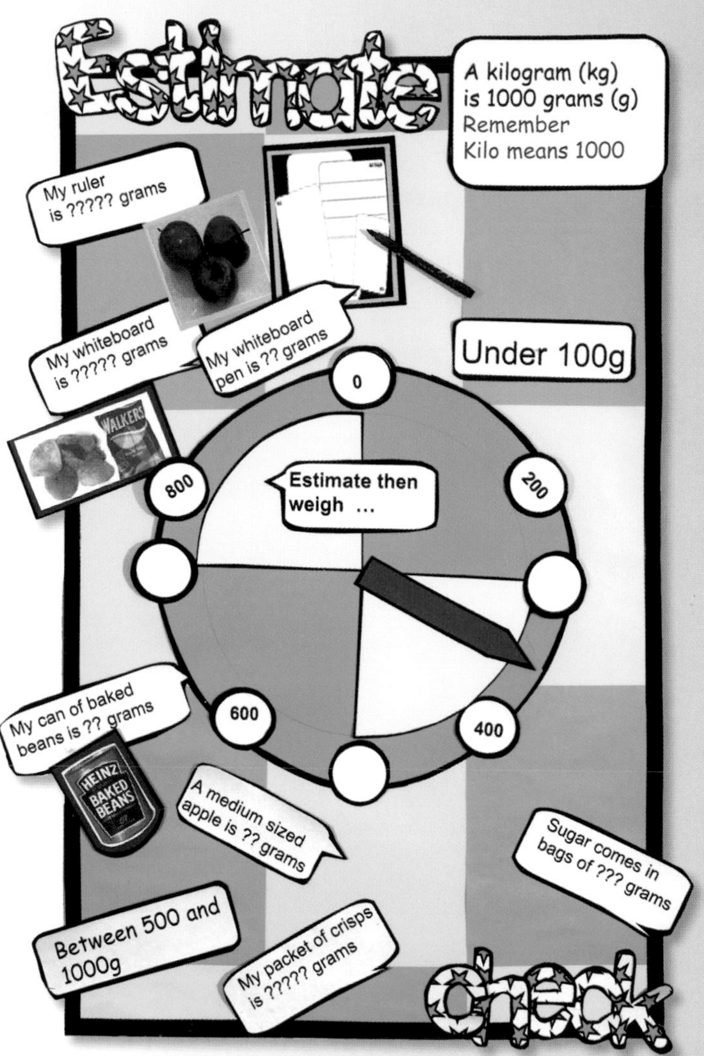

- Play *Measures Bingo*. Prepare bingo cards on A5 paper for every child, using A5 paper and a 5 × 2 table. Write a variety of answers to questions on the cards (one answer in each cell), using the key facts you want children to learn. Then ask questions: for example, 'How many grams in ¾kg?' Children then mark off 750g on their cards. The winner is the first to shout 'Bingo!' You can have the same questions on each card so they will all shout Bingo at the same time. This is also a good assessment activity to note which children didn't shout Bingo.

Further Activities

- Divide a large sheet of card into four, labelled: 'Items up to 5g', 'Items between 5 and 50g', 'Items between 50g and 500g' and 'Items between 500 and 1,000g'. Give children items to sort and estimate. They should then check and weigh them, placing them in the correct quadrant.
- Working in pairs, give children a set of scales, and fruit and vegetables (potatoes are a good option), to estimate and weigh. Ask them to estimate the weight of a piece of fruit you have chosen, then weigh and record this between them. Give children target weights: for example, 500g of apples. Children can then use a number of apples to get nearest to the weight.

Working Wall

- Produce a working wall, with a diagram of a set of balance scales, with weights balanced against familiar objects. Use Blu Tack so that different objects can be added. Children can check the weights in the display using scales.

Cross-curricular Links

- **SCIENCE** – Consider healthy eating and food values when devising and making recipes. Know what a balanced diet means.
- **PSHCE** – Keep fit and look after yourself. Eat healthily, limit your sugar input and exercise regularly.
- **LITERACY** – Interview the school cook about the healthy menus served in school.

Flying Times

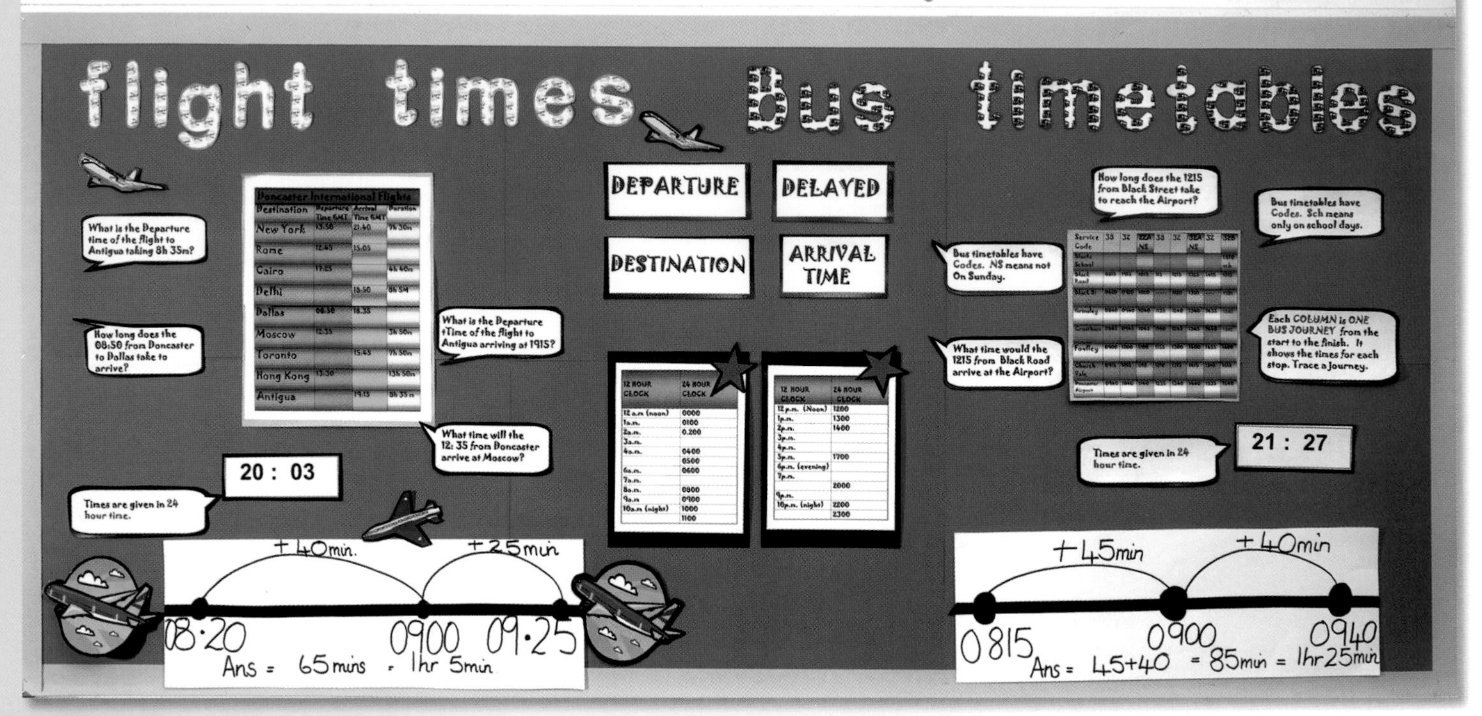

Art and Display

1. Prepare a flight timetable for an international airport (choose the airport nearest to your school). Display the airport schedule with some gaps for children to calculate the departure and arrival times, and duration of flight. Use small planes as wallpaper to make flight-times lettering.
2. Differentiate the difficulty of questions by providing timetables of shorter durations for internal flights.
3. Display a large number line showing the calculation from your local airport to Moscow. Add the destination, length of delays, and departure and arrival times.

Focus of Learning

- Reading the 24-hour clock in real-life situations

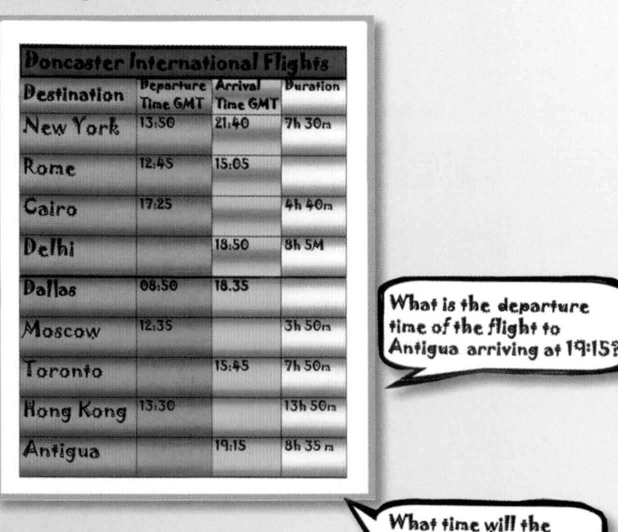

Doncaster International Flights

Destination	Departure Time GMT	Arrival Time GMT	Duration
New York	13:50	21:40	7h 30m
Rome	12:45	15:05	
Cairo	17:25		4h 40m
Delhi		18:50	8h 5M
Dallas	08:50	18.35	
Moscow	12:35		3h 50m
Toronto		15:45	7h 50m
Hong Kong	13:30		13h 50m
Antigua		19:15	8h 35 m

Starting Points

- Prepare a simple timetable showing the opening hours of the International Travel Shop times, displayed as both 12-hour and 24-hour clocks. Ask questions such as, 'What time does it close on Friday using both times? What time does it close on Sunday?' Children answer as a 'Show me' activity, using whiteboards.
- Choose a digital time: for example 20:27. Using whiteboards as a 'Show me' activity, ask children to display the time using the 12-hour clock. Ask a child to explain how to switch from a 24-hour clock to analogue or the 12-hour clock. Model the answer. For example, '20 – 12 = 8. We subtract 12 from the hours, and the minutes after the colon stay the same.' Repeat with different times. Choose examples from the departure and arrival times on the display.

- Ask children to work out the analogue time for 18.35 (when the flight from Doncaster arrives). They should explain their calculation. For example: 18 –12 = 6 hours plus 35 minutes = 6.35pm.
- To work out the flight time (the time difference between 18.35 and 8.50) children will need to use a number line to count on to find the difference. So, 18.35 – 8.50 = 9 hours 40 mins.

Further Activities

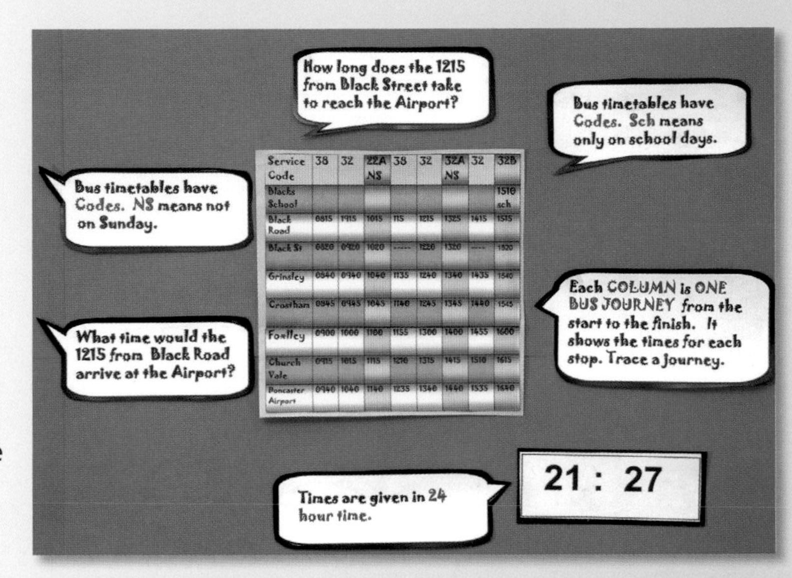

- Using the 12-hour clock, show children the time 5.42pm. Ask them to write this using a 24-hour clock time on their whiteboards, then choose a child to explain what this time would be. Model the child's answer. For example, 'Because the 24-hour clock continues from 12, we need to add on two hours, so 5.42pm is 17.42.'
- Ask children to answer questions such as, 'What's the flight time of the 08:50 from Doncaster airport to Dallas airport?' They should answer the question using a number line.
- Prepare a bus timetable with terminal times at your nearest international airport. Each column is one bus journey from the start to the finish, with times shown at each stop. Explain that individual columns have been highlighted to show specific journeys. Emphasise that codes are important when reading timetables. For example, people need to know if the bus runs daily: NS means not on Sunday; Sch means that the bus only runs on school days.

- Provide a table-top display with two clocks for children to use to show the arrival and departure times of aeroplanes. Laminate number line frames for children to use, and add flight times questions for children to answer.

Working Wall

- Prepare a laminated table showing conversions for the 12- and 24-hour clocks. Children can then fill in the missing times and use the table as a reference.

12 HOUR CLOCK	24 HOUR CLOCK
12 a.m (noon)	0000
1a.m.	0100
2a.m.	0.200
3a.m.	
4a.m.	0400
	0500
6a.m.	0600
7a.m.	
8a.m.	0800
9a.m.	0900
10a.m. (morning)	1000
	1100

12 HOUR CLOCK	24 HOUR CLOCK
12 p.m. (Noon)	1200
1p.m.	1300
2p.m.	1400
3p.m.	
4p.m.	
5p.m.	1700
6p.m. (evening)	
7p.m.	
	2000
9p.m.	
10p.m. (night)	2200
	2300

Cross-curricular Links

- **GEOGRAPHY** – Plan imaginary holidays, work out flight times, read a timetable, and work out departures and arrival times. It is also useful to use real bus timetables for the area in which children live, for planning journeys.

Malicious Measures

Focus of Learning

- Developing vocabulary and recall of measures facts

Art and Display

1. Use a theme that supports your class topic. In this example, a 'spooky' theme was used that enhanced the literacy topic of Macbeth. Other possible themes are Sports Day or Healthy Eating.
2. Use Clip Art to create a number of related images. For each image, pose a problem that makes children focus on measures. For example, 'If one cockroach weighs 5g, how much do six cockroaches weigh? A frog jumps 16cm. How far is this in millimetres?'
3. Add examples of resources that support measures, such as a metre stick, a negative number line and a protractor. Add images appropriate to the topic.

Starting Points

- Ask children to write a short estimation story. For example, 'I woke up this morning and looked at the clock. I had a bath, then I ran downstairs and made myself a bowl of cereal.' Give children a whiteboard and pen and ask them to write the number of measures to be recorded, and record the measure next to its number. They can choose the correct measures from the teacher's board: for example, ml, l, cm, m, g, kg, seconds, min, hr, then record them. Then ask them to write their own estimation stories with illustrations for the class and/or partners to answer. A fun activity is for them to choose an appropriate Clip Art image and write their story for the book using Microsoft Word. Then make these stories into a class book for children to refer to.

Further Activities

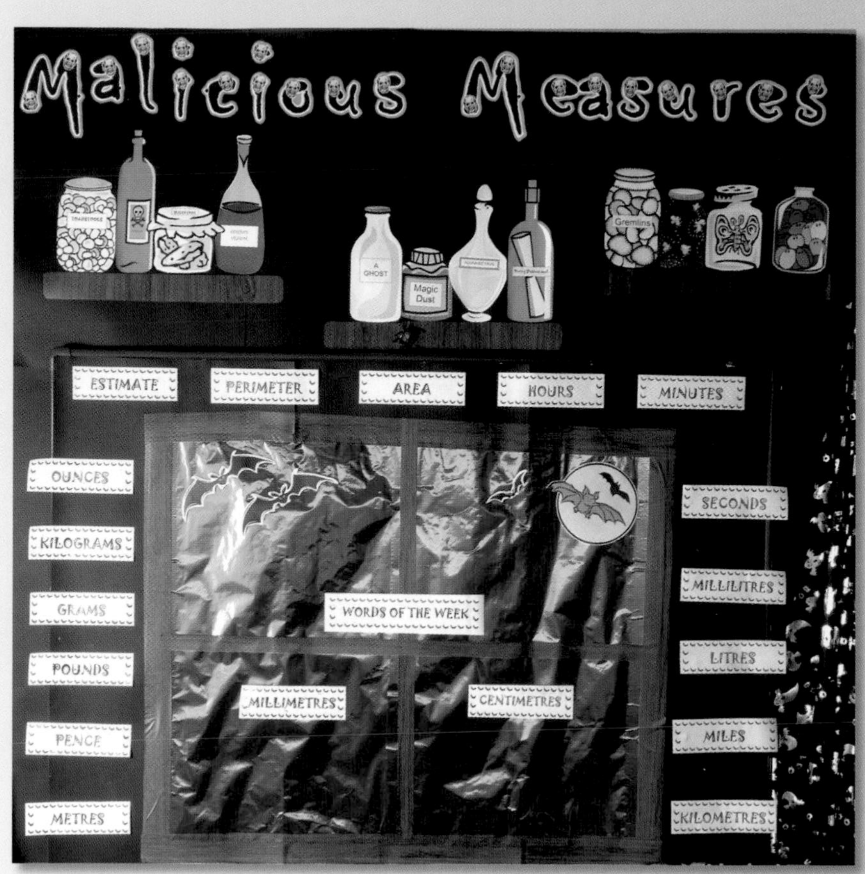

- Give groups of four children a variety of different lengths of ribbon, material, sequins, wood and wool. Ask them to estimate if they are more or less than 1m, 50cm, 1.5m, and so on. Ask them to measure accurately and write the difference between their measurements and estimations. You can set up similar activities for weight, liquid measure and time.
- Make a range of whole-class bingo cards using a 6 × 2 grid displaying different measures: for example, l.5l, 500ml, 50cm, one-and-a-half hours, 45 minutes, 1 500g. Call out the measure 1 500ml and children cover 1.5l.
- When considering measures, it is essential to provide practical activities using standard measures, rather than just using interactive teaching programmes. Give children topic-linked objects to weigh, with problems for them to solve. For example, 'Meg the witch wants to make a potato soup. She has 350g of potatoes. How much more does she need to make 1kg of potatoes?' Give children ten potatoes. Then ask, 'How much does the heaviest potato weight? How much does the lightest weigh? What's the difference in weight between them?'

Working Wall

- Design a window frame, splitting it into four equal parts. Then make and display a set of cards using the vocabulary of measures, as displayed. Children should choose a measure of the week and attach it to the window. Ask children to write down on their whiteboards as many objects as they can that would be measured in, for example, mm or kg. They can choose Clip Art or create their own drawings and attach them to the working wall to illustrate the measure and related vocabulary. Use vocabulary from the working wall for children to practice spelling for homework.

Cross-curricular Links

- SCIENCE – Children can collect a variety of twigs, leaves and tree trunks. Then they can estimate their length and weight in cm, m, kg and g, and check using the appropriate measure.

  – Plan measures into sports day: length of races, jumps, time taken to run races, liquid required for drinks, weight of biscuits for refreshments, and so on.
- ART – Make webs and spiders to illustrate the Malicious Measures display.

Feeding Time at the Zoo

Focus of Learning

- Using timetables to support work on time differences

Art and Display

1. Create a Feeding Time at the Zoo timetable. Ensure the information can easily be removed to allow the display to be used over an extended period of time.
2. Create pictures of the appropriate zoo animals using Clip Art or artwork.
3. Generate and display questions relating to time differences.
4. Create a large snake, labelled 0–60, to be used as a number line for time number line calculations.
5. Include a clock to support children in developing an understanding of time passing.

Starting Points

- Prepare a counting stick labelled 0–60 and ask children to suggest why it's labelled like this. Add labels showing 30 minutes, half an hour, 15 minutes, quarter of an hour, 45 minutes, three-quarters of an hour. Explain that the clock face is a curved number line.
- Give children intersecting circles of two colours. Ask them to estimate fractions of time: for example, 'Show me 15 minutes/quarter of an hour.' Children should reply showing a quarter of an hour by moving 90 degrees or a quarter turn (the quarter turn will be highlighted in one colour showing one quarter of the circle).
- Use the snake number line to find time differences. For example, 'We arrived at the zoo at 10.30am and met in the café to eat our lunch at 1.30pm. How long did we have to wait until lunch?'

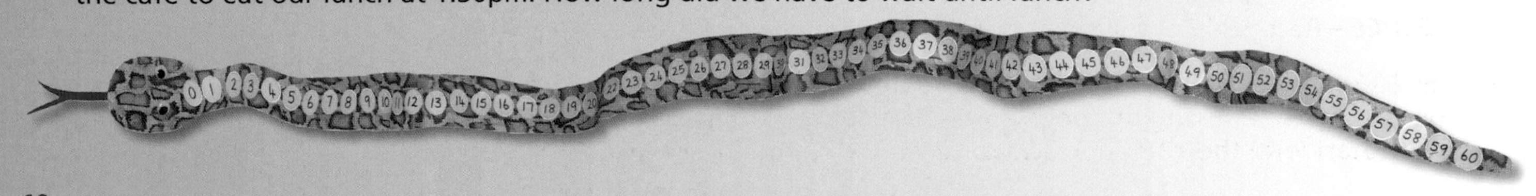

Further Activities

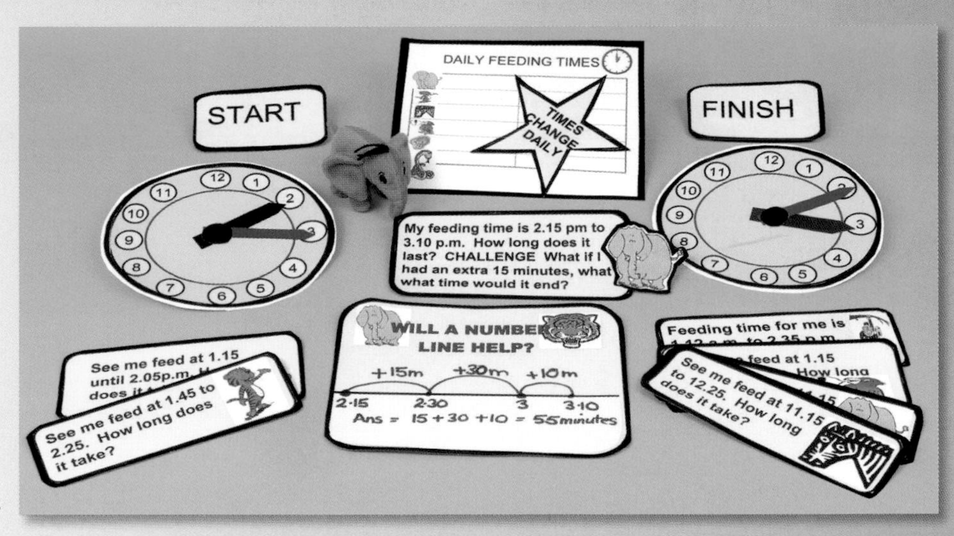

- Prepare a table-top display with 12-hour clocks to support children in answering time difference questions, and to model number line frames as a prompt for children to work from. Add examples of illustrated problem cards. Add a Daily Feeding Times timetable, which you can change daily to vary the difficulty of the calculations.
- Give a child the class watch, and they become 'keeper of the time' for a day, reminding you and other children about important times and the passage of time. Encourage them to express times as both analogue and digital.

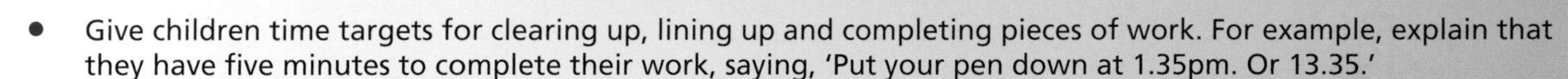

- Give children time targets for clearing up, lining up and completing pieces of work. For example, explain that they have five minutes to complete their work, saying, 'Put your pen down at 1.35pm. Or 13.35.'

Working Wall

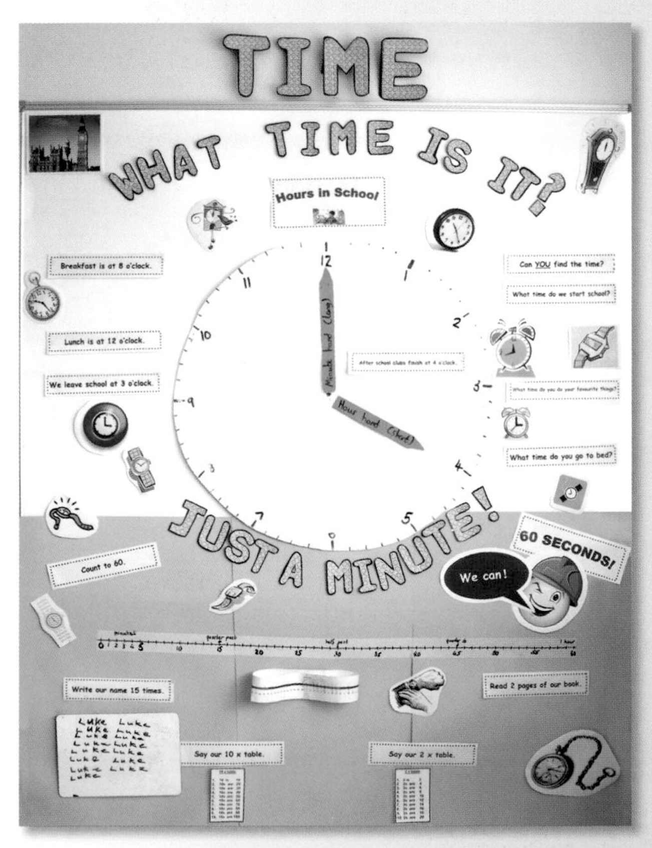

- Prepare a time working wall, divided into two. On the top half, design a large 12-hour clock, with the hands labelled as 'Hour' and 'Minutes'. Set the clock at important times for children: for example, when break or afternoon clubs start. Surround the clock with questions appropriate to children's lives. For example, 'What time do we start school? What time do you go to bed?' Use the display to ask questions such as, 'What can we do in five minutes/half an hour?', and so on. Add questions involving half and quarter hours, once children understand how to read the hours. On the bottom half of the display, add a label '60 seconds', and add all the things that children can do in that time. For example, 'Say our 2 × table', 'Write our name 15 times', 'Read two pages of our book'. Display a number line showing 60 seconds in one minute.
- Involve children in planning day trips, and encourage older children to plan start and finish times on sports day. Children often enjoy deciding on the winners of a race by ordering their times. This working wall could easily be changed to a simpler race, where the times are just in minutes.

Cross-curricular Links

- **PE** – Use stop watches to record timings of races. Record time for games such as football, netball, and so on.
- **SCIENCE** – Record times for recovery rates after exercise, and so on.
- **GENERAL** – Focus on time throughout the curriculum by giving start and completion times and targets for completing work, lining up for assembly, and so on. Consistently inform children about time that is relevant to them. Start with the 12-hour clock and move on to the 24-hour clock.

Pirate-Packing Problems

Art and Display

1. Design and make a pirate and add a pirates' treasure chest. Use skull and crossbones Clip Art as wallpaper to make the lettering and the pirate ships, adding lollipop sticks for sails.
2. Ask, 'How many golden bars measuring 1cm^3 will fit into the boxes to store the pirates' gold?'
3. Using the pirate wallpaper, children can wrap a variety of boxes, measure them and add labels with their length, width and height. They can then calculate their volume.

Focus of Learning

- Finding the volume of cubes and cuboids

Starting Points

- Discuss the importance of knowing how much space is available in a cube. Give pairs of children 20cm cubes. Ask them to make as many different cuboids as possible. Children can write the calculation of length × breadth × height on their whiteboards and display (you may prefer to use the term 'width' rather than 'breadth'). Explain that the amount of space that is taken up by each cuboid is its volume: $5 \times 2 \times 2 = 20$cm^3.
- Prepare different examples of cuboids using centimetre cubes, then ask children to calculate on their whiteboards the number of cubes used. Alternatively, give examples for pairs of children to calculate.
- Explain the difference between cubes and cuboids. A cube must have six square faces, 12 edges and eight vertices: for example, a dice. A cuboid has six rectangular faces: for example, a cereal packet.

Further Activities

- Set up the following investigation for children. Using centimetre cubes, build three cuboids. A measures 2cm × 2cm × 2cm; B measures 2cm × 2cm × 2cm; C measures 2cm × 2cm × 8cm. Ask children to design and make a treasure chest that Pirate Pete's treasure will fit into exactly (this box should measure 2cm by 2cm by 4cm). Ask them, 'Could the treasure chest have had different dimensions and still be the correct size to hold the treasure?' Explain that Pirate Bill needs to have an open box to store 36 gold cm cubes. Ask children to design and make a box to store the cubes in. They should make a label explaining how they designed their box and stating its cubic capacity. Encourage children to use the cubes to check their design.
- As a final challenge, explain that there were no boxes left for the last of Pirate Bill's gold, which measures 3cm × 3cm × 3cm. He decided to paint the gold red on the outside. Ask children to work out the following: 'How many cubes have no faces painted? How many cubes have paint on one face only? Can you work out how many cubes will have two, three or four faces painted?'
- Choose a variety of boxes and wrap them as presents. Ask pairs of children to investigate how many gold cm cubes would fit into the boxes. Children can then label the boxes and record their calculations.
- Challenge children to make Russian-doll style nesting boxes, decorated with pirates and pirate ships (see display). Give groups of four children cm^2 paper and ask them to make one open cube that measures 10cm × 10cm. Each child can then make a different-sized open box – each one $1cm^2$ smaller than the next – to fit into the 10cm × 10cm box. Give children a net for making the dolls, which they must alter for each size. More elaborate dolls can be made using card.

Working Wall

- Challenge children to decide how many cubes could fit in the boxes, reminding them that the space taken up by a 3D shape is called volume. Volume can be measured by counting the number of cubic centimetres that fit inside a shape. Invite children to explain how many more cubes are needed to fill these boxes. Ask them to make boxes of their own and work out how many cubes will fit into the boxes. More able children could design and make cubes and cuboids that are a different shape but have the same volume.
- When the children have designed, made and decorated their boxes to hold the cubes, invite them to write their own questions relating to the volume. For example, 'How many more cubes do I need to fill this box?' Attach the boxes to a working wall for members of the class to answer the problem.

Cross-curricular Links

- **DESIGN AND TECHNOLOGY** – Make and design nets and boxes for a purpose. For example, make toothpaste packets or design a sweet packet.
- **ART** – Make padded material portraits of pirates. Large speech bubbles can be added and displayed next to the portrait. For example, 'I have 48 gold cubes. Design and decorate a box for my treasure.'

Maths through Art (Length)

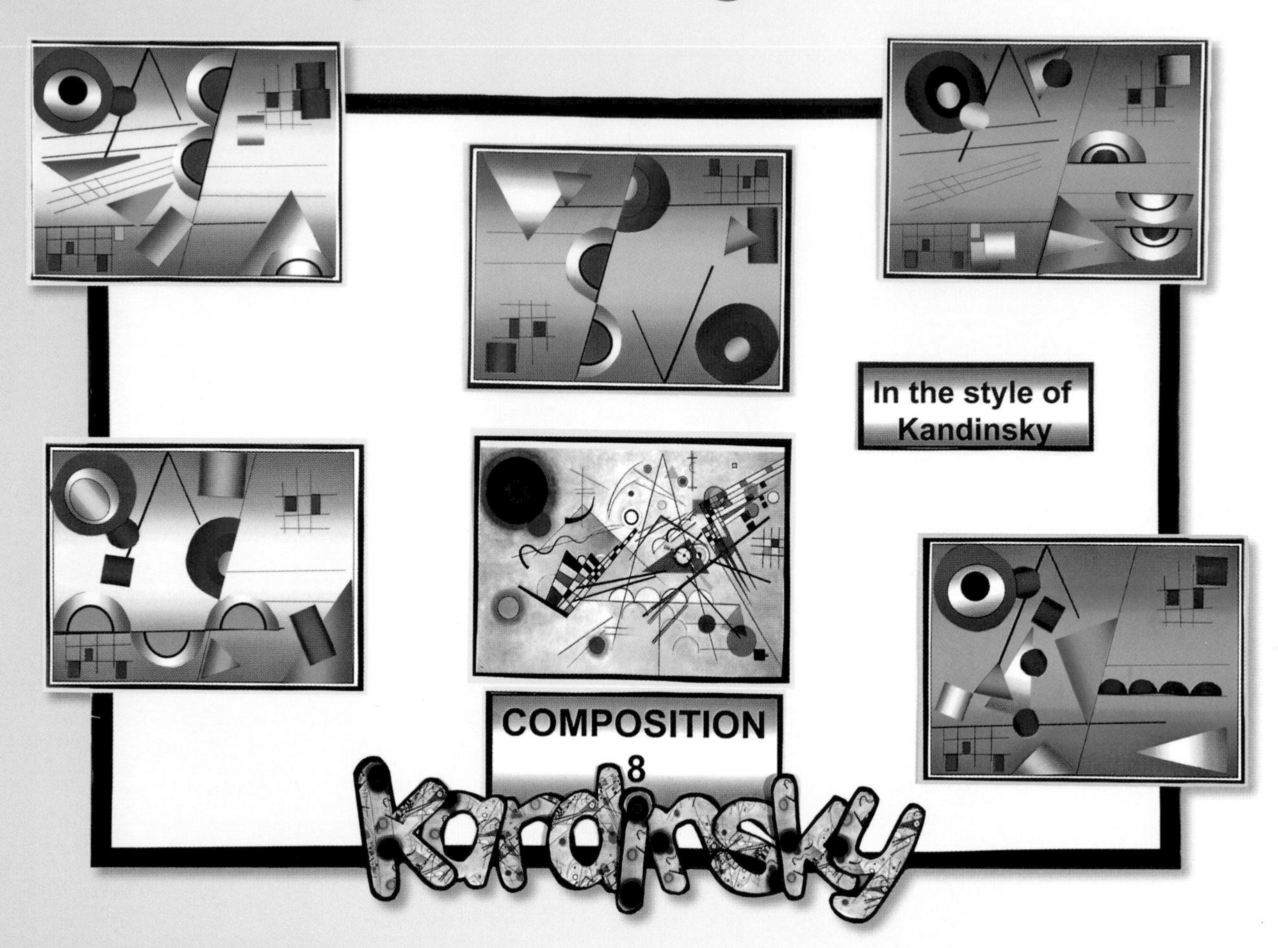

Art and Display

1. Studying the work of Wassily Kandinsky (1866–1944), in particular *Composition 8*, can support an understanding of describing and classifying shapes, and its practical application. Discuss how shape has been used in *Composition 8*, and ask children to describe and classify the shapes that they have seen.
2. Give younger children templates of quadrilaterals, semi-circles, polygons that are both regular and irregular, and triangles. These can be used to cut out the desired shape and make art in the style of Kandinsky. Alternatively, children can use software such as AutoShapes in Microsoft XP to make the shapes. The shade tool gives a luminous quality to the artwork. Straight lines can be drawn using either the line tool or a black pen. More able children can use sets of compasses, and protractors and rules, to construct their own shapes. Although working individually supports children in their mathematical knowledge, working in groups of four encourages and supports mathematical discussion. This is a very simple and effective activity that produces high-quality work.

Focus of Learning

- Understanding and investigating length, horizontal and vertical lines, and area and shapes

Starting Points

- Give children the opportunity to study and produce work in the style of these artists and, through this practical activity, to develop their understanding of describing and classifying shapes, and repeating patterns.

Further Activities

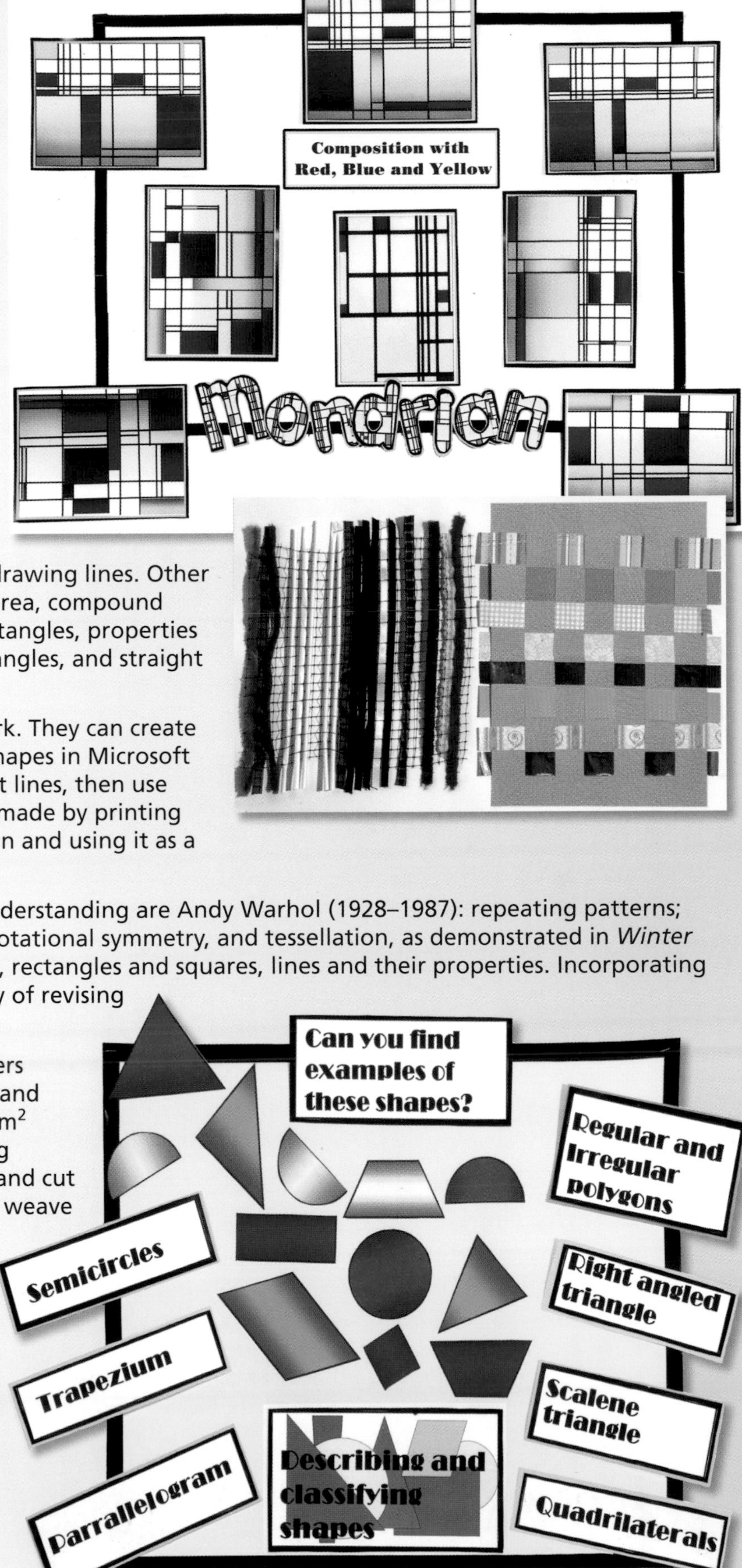

- Discuss the shapes and properties that can be seen in *Composition with Red, Blue and Yellow* by Piet Mondrian (1872–1944). Give groups of four children a metre square sheet of card or cheap canvas. Ask them to measure the card using a metre stick, then draw a straight line from the top to the bottom (use the word 'vertical'), then measure the breadth/width of the card (use the word 'horizontal'). Ask children to add up to ten more lines, either vertical or horizontal, that make rectangles and squares. Point out that a vertical line forming a right angle with the horizontal line is perpendicular. Using three colours only, children can work on different parts of the painting, but make sure that no shapes touch the same colour. Ensure that children leave white spaces, as Mondrian gave texture to his spaces by drawing lines. Other mathematical concepts can be: perimeter, area, compound area, surface, properties of squares and rectangles, properties of lines (parallel and perpendicular), right angles, and straight lines.
- Older children can use IT to produce artwork. They can create work in the style of Mondrian using AutoShapes in Microsoft XP and the line tool. They can draw straight lines, then use the fill tool with shading. Lettering can be made by printing one of the artworks in the style of Mondrian and using it as a background.

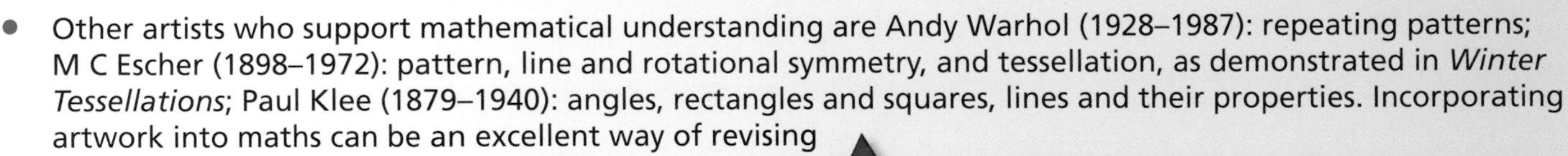

- Other artists who support mathematical understanding are Andy Warhol (1928–1987): repeating patterns; M C Escher (1898–1972): pattern, line and rotational symmetry, and tessellation, as demonstrated in *Winter Tessellations*; Paul Klee (1879–1940): angles, rectangles and squares, lines and their properties. Incorporating artwork into maths can be an excellent way of revising and practically applying these concepts.
- Children could measure and cut pipe cleaners and art straws to 15cm long with accuracy, and arrange them vertically and in parallel, in cm^2 netting. Alternatively, a card paper weaving frame can be made. Children can measure and cut weaving materials measured at 12cm, then weave them into the frame. They should use the terms 'vertical', 'horizontal' and 'parallel' to describe their weaving.

Working Wall

- Shapes can be made using AutoShapes and the shading tool to create a working wall. Challenge children to identify the shapes, and their properties, as displayed on the working walls in the work of Kandinsky and Mondrian.

Sweet Data (Charts and Tables)

Art and Display

1. Create an empty graph with a title and labelled X and Y axis. Using a packet of sweets to demonstrate, fill in the data on the graph relating to the number of sweets as shown in the display.
2. Invite children to sit in pairs and write four questions and answers relating to the graph. For example, 'Which sweet occurs most often? The purple sweet, because there are eight, which is one more than the number of red sweets, which is seven.' Give children a 'because' card each to help them structure their answers.
3. Explain that the information in the table shows the number of sweets that three teachers had in their packets. Ask children if they have enough information to answer the questions on the display. For example, 'There are the same number of sweets in every packet. True or false?' or 'The factory makes the same number of each colour of sweet. True or false?' Children should justify their answers using the word 'because'.
4. Ask probing questions such as, 'What if each step represented 7 sweets not 1? How many blue sweets would Mrs Share have?'

Focus of Learning

- Interpreting charts and tables

What if each step represented 7 sweets not 1? How many blue sweets would Mrs Share have?

Colour	Mrs Share	Mr Burton	Mrs Jones
Blue	6	3	4
Brown	5	5	5
Green	4	4	4
Orange	3	6	6
Pink	6	6	5
Purple	8	9	7
Red	7	6	7
Yellow	4	4	5

Starting Points

- In the hall or playground, give each child a coloured counter. Then ask them the following questions. 'How many red, green, yellow, brown, orange, pink and blue counters are there?' Ask the children to order themselves in rows according to the colour of their counters. As a class, record the information using a tally chart, then transfer this onto a frequency chart. You could also challenge more able children to find the mean, mode and range of their information.
- Create a graph without a title or labels on the X and Y axis. Ask children to hypothesise what the graph could be displaying and why. Ask, 'What information do you already know about the graph?' For example, the number of columns and the smallest/largest amount. 'What information do you need to make sense of the graph?' For example, a title explaining what the graph is about, the number of sweets to record on the Y axis, and the different colours of sweets to record on the X axis. Give younger children a proforma to arrange the number and colours of counters.

Further Activities

- Ask children to work in groups of six. Give them small boxes of Smarties (boxes or tubes containing coloured counters can be equally effective). Ask them to make a tally chart showing the total number of sweets in each packet and the total number of each sweet. When you say 'Show me', they show you their recordings and you discuss their appropriateness. Ask, 'Which was the easiest tally chart to understand and why?' Then compile a list showing the information gathered from each group. Using the data collected, children can answer the following questions using their whiteboards: 'Did each packet have the same number of sweets? How do you know? Explain your calculation. Was there the same number of each colour of sweet in the packets? How do you know? Explain your calculation.' A further challenge could be to find the mean, range and mode of the data.
- Pose the following problem for more able children. Explain that there are the same number of each colour of Smarties in each packet. From the information that they have been given, children should produce a tally chart and a graph. They could also find the mode, mean and range of the data.

Working Wall

- Produce a working wall that presents a graph with data missing. Add the labels X, Y, Axis, Title, Labels, Steps. Challenge children to put the labels in the correct position on the graph. Then ask them to write their own questions relating to the data, using Post-it notes to add them to the display.

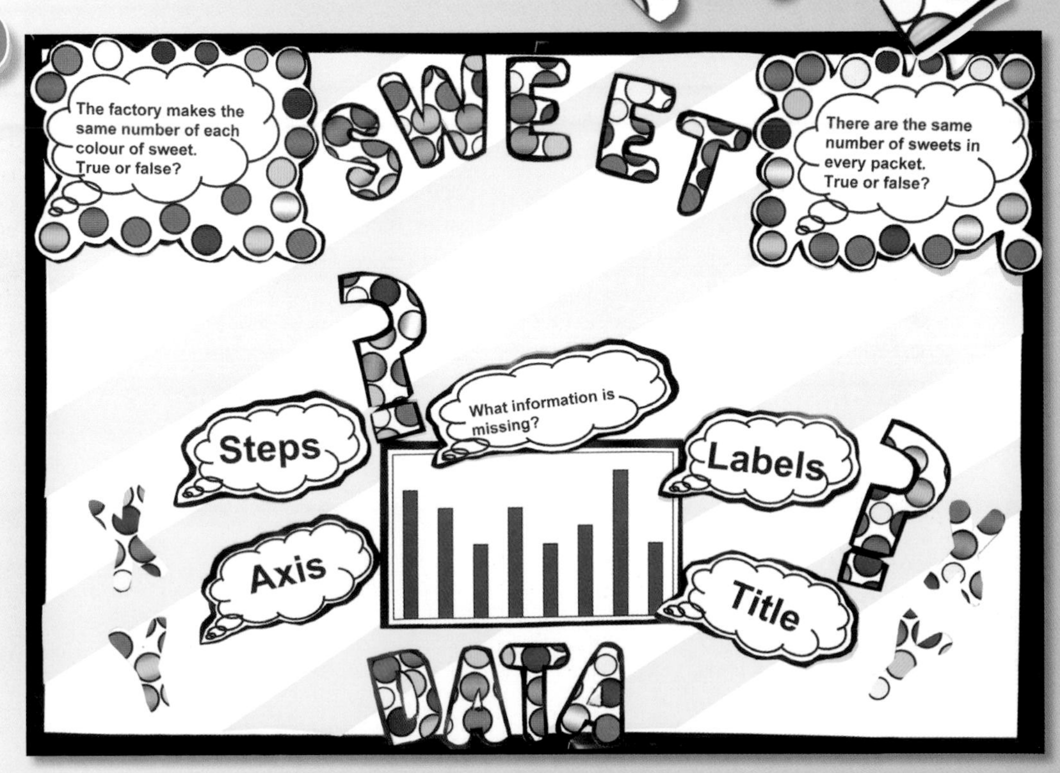

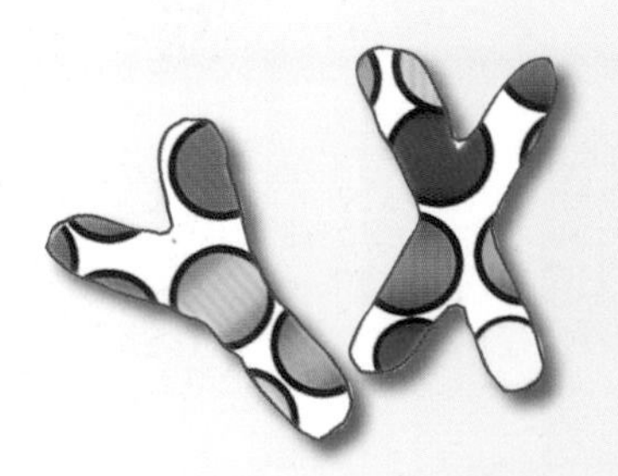

Dinosaurs (Diagrams)

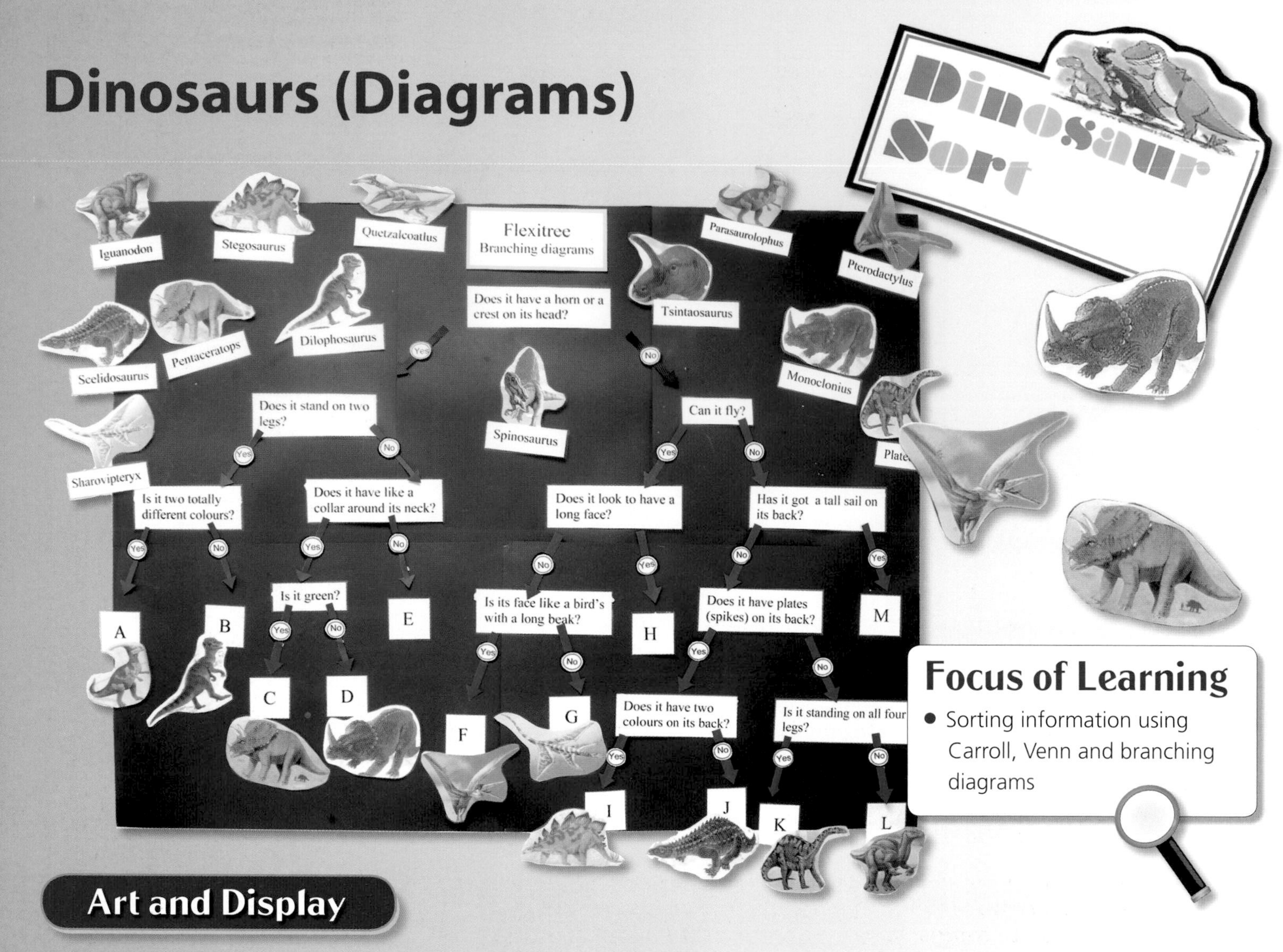

Focus of Learning

- Sorting information using Carroll, Venn and branching diagrams

Art and Display

1. Create a flexitree branching diagram to sort the different features of dinosaurs. For example, 'Does it stand on two legs?' Sort according to the agreed criteria. Add arrows labelled 'Yes' or 'No' to show the route of the dinosaur. Use Clip Art of dinosaurs to illustrate specific features.
2. Provide children with Clip Art of the named dinosaur as a reference as they sort the dinosaurs.

Starting Points

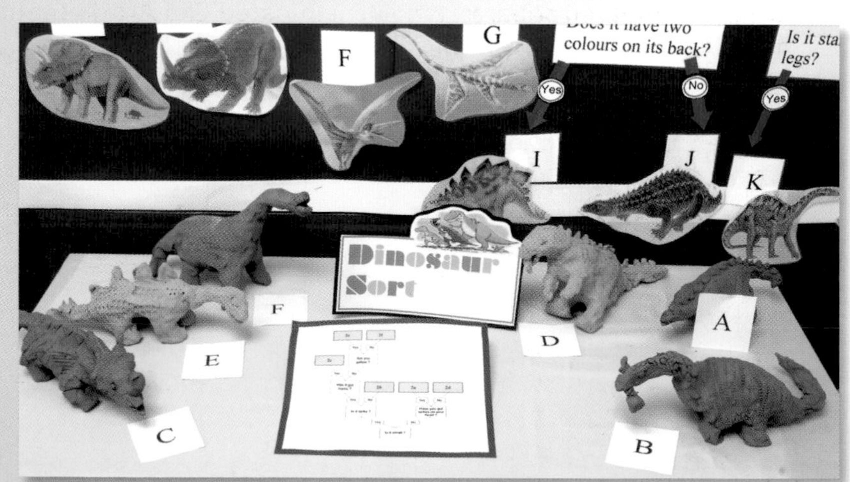

- For work on branching diagrams, present children with a group of animals/dinosaurs and ask them to sort according to the agreed criteria (for example, number of legs), then divide the animals into subsets: has horns, has no horns, and so on.
- For work on Venn diagrams, draw a circle on the floor and ask children who have a specific criteria (for example, are seven years old) to stand in the circle. Repeat with different criteria.
- Play the *Yes or No Game* as a starting point for work on Venn diagrams. Ask children to pick a number between 1 and 20. You need to have already chosen a criterion (for example, multiples of 3), then say yes or no depending on whether their choice matches this. Write numbers that don't meet the criteria outside the circle. The child who guesses the criteria first is the winner. Make the activity more challenging by using more difficult tables.

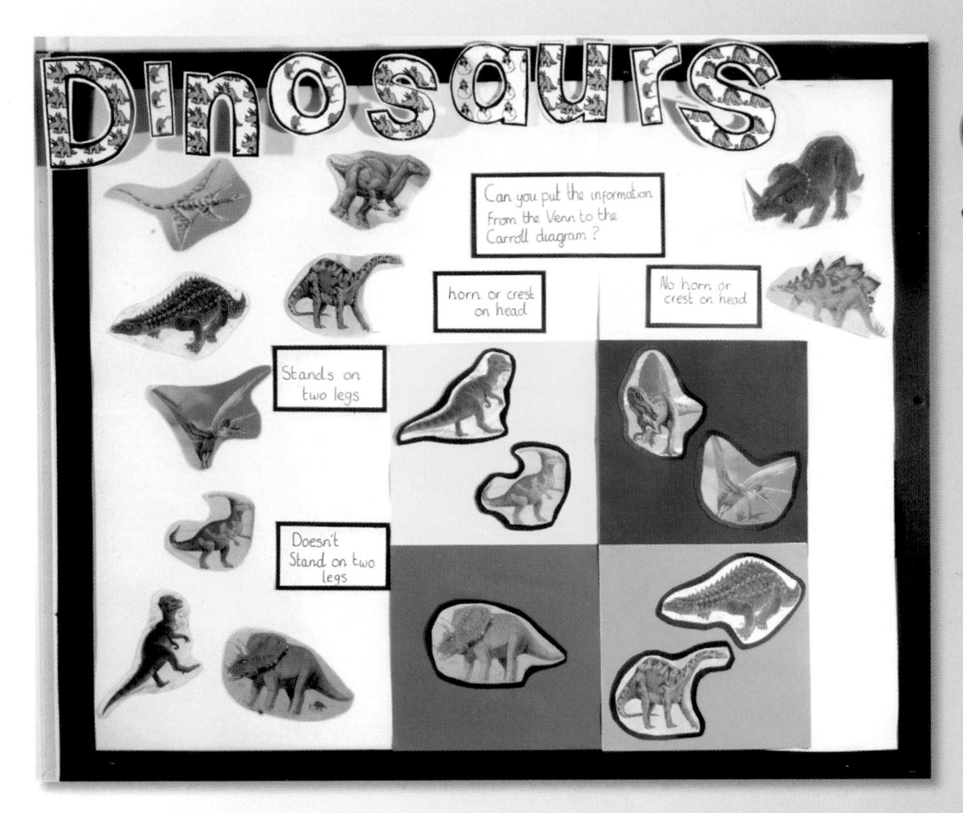

Further Activities

- For work on Carroll diagrams, use the hall or playground as a working space. Explain that the space is divided into quadrants, which represent a Carroll diagram. Give children cards with numbers up to 50, and ask them to stand in the appropriate quadrant. Use this idea to sort children, for example, with/without brown eyes, and those with/without brown hair. If you don't have space to create a Carroll diagram, draw a large one instead and ask children to write their names and criteria, and put them in the appropriate place.

- For work on Venn diagrams, use two PE hoops: one labelled 'straight sides' and the other 'curved sides', with an intersection for shapes with curved and straight sides. Give each child a capital letter and ask them to put their letter in the appropriate place. In pairs, ask children to generate questions relating to the diagram on whiteboards. For example, 'Where would we put B because it has curves and straight lines?' Play *Spot the Mistake*: put a capital letter in the wrong section and ask children to explain why it is incorrectly placed.

- Use hoops labelled with the properties of 2D shapes. Give children a card showing a specific 2D shape. Then children can put their cards in the correct place on the Venn diagram. The challenge is for children to sort according to three criteria; make sure there are intersections between the criteria.

Working Wall

- Produce a working wall, using Clip Art images as wallpaper for lettering. Make a Carroll diagram using different coloured squares. Provide laminated Clip Art dinosaurs. As a class, sort the dinosaurs according to named criteria and, over time, pick pairs of children to sort a variety of dinosaurs. Change the criteria so that children can experience sorting in a different way.

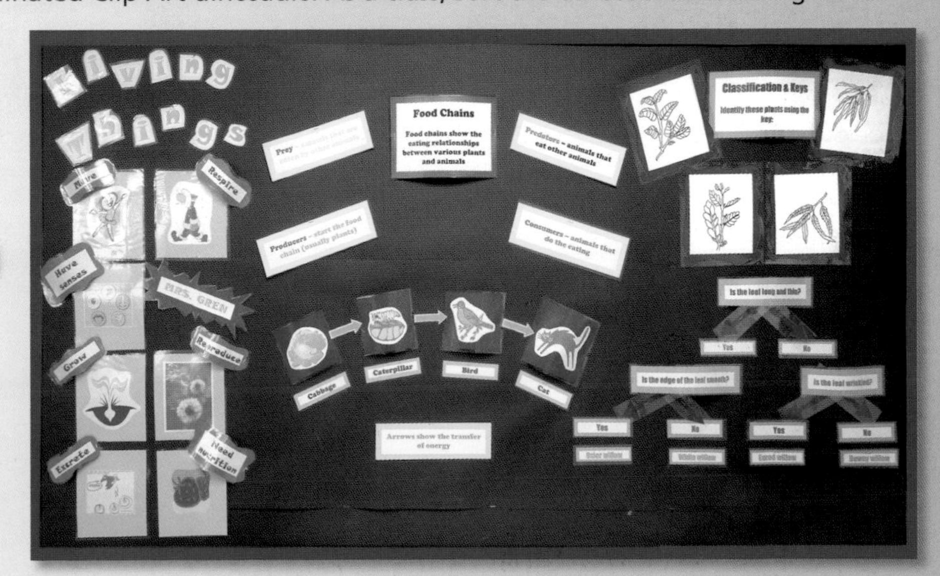

Cross-curricular Links

- **SCIENCE** – Sort features of animals and plants into branching diagrams with appropriate questions that can be answered using 'Yes' or 'No'.

- **ART** – Make clay dinosaurs to use for sorting according to features. Make observational drawings of dinosaurs to highlight features to be sorted.

Keeping Up with the Kids

Art and Display — Multiplication and Division

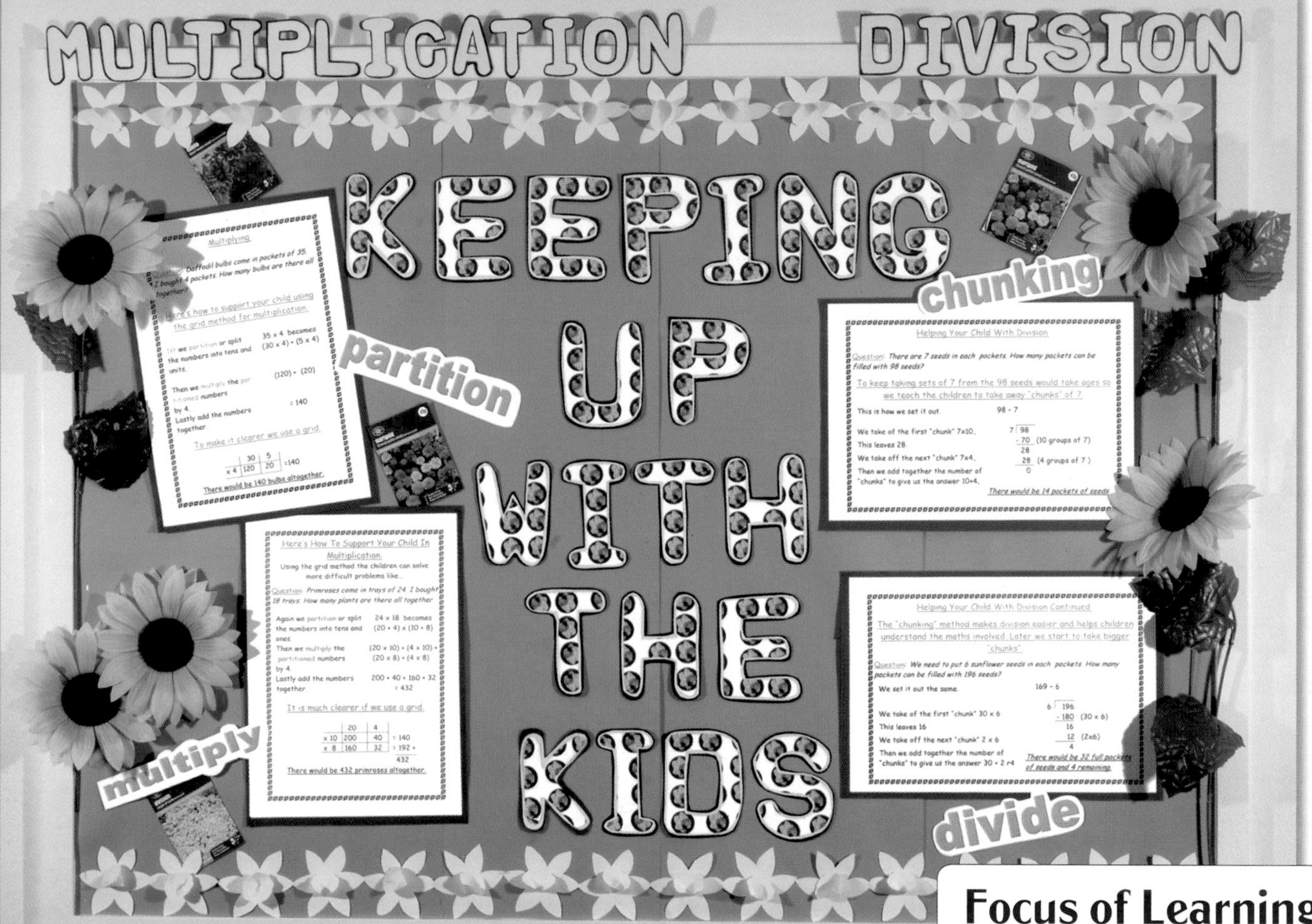

Focus of Learning

- Displaying your school's calculation methods to support parental involvement with written calculations

1. Create a garden-centre problem solving display that illustrates your school's calculation policy for multiplication and division, using partitioning, multiplication and division. Display essential vocabulary, including multiplication, division, chunking and partitioning.
2. Decorate using sunflowers and seed packets, and make small paper daffodils as a border pattern. Add the relevant vocabulary: chunking, divide, partition and multiply.
3. Create a step-by-step set of instructions for multiplication and division problems and display them. Note in the above example that the calculation 7 × 10 would be read as 7 multiplied by 10, which is seven, ten times.
4. Provide copies of these examples to support parents and children at home. It is useful to use A4 paper, cut in half and laminated, so they can be used again.
5. Display the homework information for parents to take home and use as a reference.

Art and Display — Addition and Subtraction

1. Continue the garden-centre theme with a problem solving display. This should demonstrate addition and subtraction, and counting up to find a difference with a number line. Use number lines for calculating time differences and measures.
2. As part of the garden-centre theme, use simple frames as shown in the display to support children in answering problems using number lines and to inform parents about calculation methods.
3. Include the relevant vocabulary on vocabulary flowers.
4. It is useful to laminate the questions and send them home to support parents and children in using number lines in solving problems.

Further Activities

- Create a table-top display, showing your school's agreed calculation policy, open at the entries on number lines where step-by-step instructions are written and illustrated with examples. If you don't have a policy, give examples of the methods you use in the context of the garden centre. Add a selection of the apparatus that you use in school to support children using number lines. Show examples of children's books and work using the displayed calculations: these can be photocopied and displayed or placed in plastic wallets. Once you have an agreed policy, it is useful to give children examples of the calculations as models, and print these onto sticky labels that can be stuck in books to support children.

Working Walls

Working walls are designed to support learning and teaching in a lesson or series of lessons. They display the main objectives of the lesson visually and kinaesthetically, with special emphasis on the development of mathematical vocabulary. Where appropriate, interactive activities can be displayed for children to use: for example, numbers to order, tables to fill or multi-link cubes to count.

Position working walls at the front of the teaching area so that you can refer to them during the lesson. Also, children will know where to look for these prompts to support independent learning and self-confidence. They are simple and uncluttered, and need little preparation. They can display your school's preferred method of calculation: for example, how to use number lines, problem solving strategies and examples of co-ordinates – and are relevant to all areas of the numeracy curriculum.

They should display the relevant vocabulary and 'tips', to support children with challenging points in numeracy. For example, children need to know that when using a number to count up to find a difference, the answer will be in the jumps above the line (many children become confused and don't know where to find the answer).

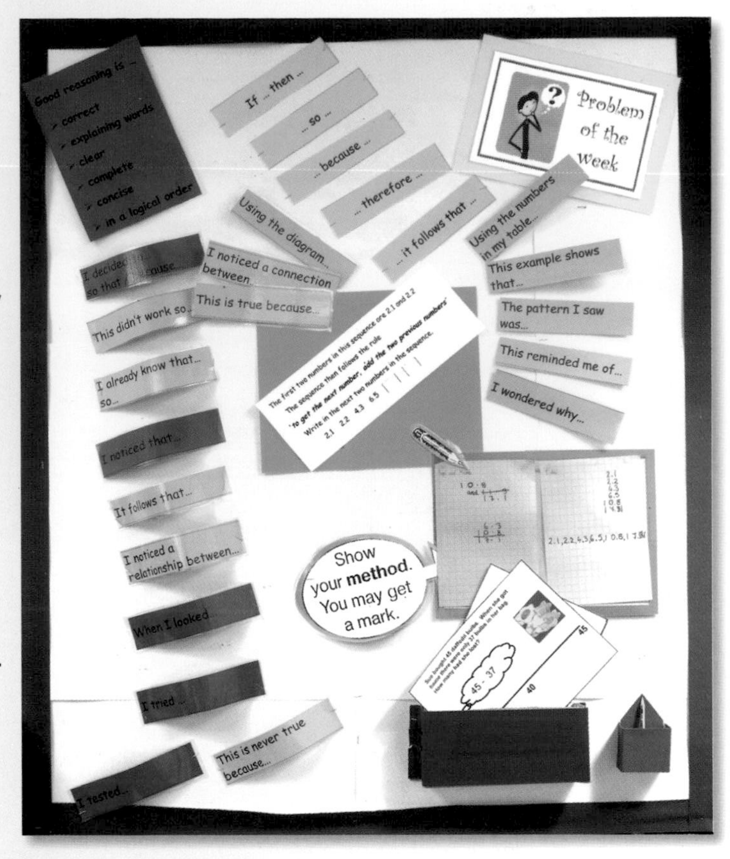

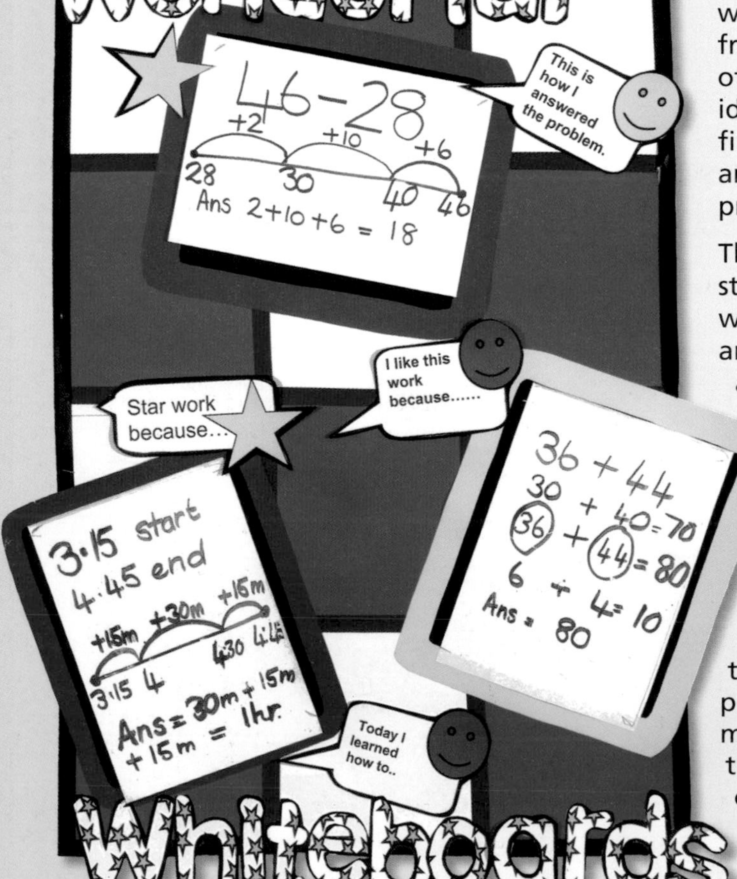

Whiteboard work is an essential part of interactive teaching and learning; however, excellent work is often wiped away. It is useful to photocopy examples of whiteboard work and display it. You could provide simple frames to lift its status, and add the reasons why a piece of work is 'good', highlighting it or asking a child to identify and write why their work was chosen. You may find it useful to move the material on the working wall to another area and display it for an extra week to recap on previous work.

The displays can be stored in A3 plastic wallets, clearly labelled and used many times, as they support the key objectives in the numeracy strategy. In a survey, children were asked what they found helpful with their maths in the classroom: 90 per cent of them mentioned that the information on working walls reminded them of what the teacher had said so they could 'get on with their work on their own'.

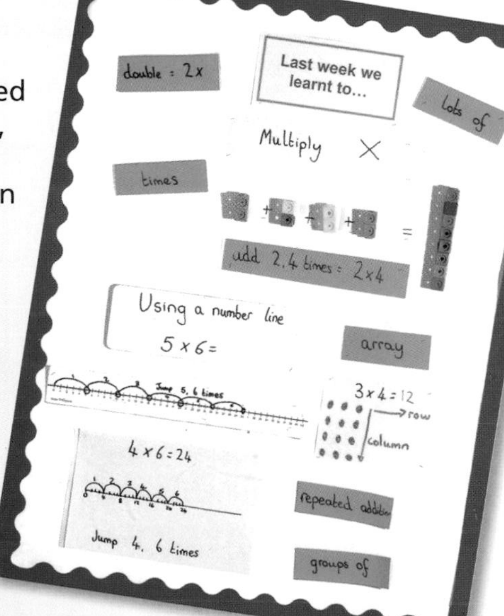